FREE Test Taking Tips DVD Offer

To help us better serve you, we have developed a Test Taking Tips DVD that we would like to give you for FREE. **This DVD covers world-class test taking tips that you can use to be even more successful when you are taking your test.**

All that we ask is that you email us your feedback about your study guide. Please let us know what you thought about it – whether that is good, bad or indifferent.

To get your **FREE Test Taking Tips DVD**, email freedvd@studyguideteam.com with "FREE DVD" in the subject line and the following information in the body of the email:

a. The title of your study guide.

b. Your product rating on a scale of 1-5, with 5 being the highest rating.

c. Your feedback about the study guide. What did you think of it?

d. Your full name and shipping address to send your free DVD.

If you have any questions or concerns, please don't hesitate to contact us at freedvd@studyguideteam.com.

Thanks again!

GMAT Prep Book 2022 and 2023

GMAT Study Guide with Practice Test Questions Review [7th Edition]

Joshua Rueda

Written and edited by TPB Publishing.

ISBN 13: 9781637753828
ISBN 10: 1637753829

Table of Contents

Quick Overview

As you draw closer to taking your exam, effective preparation becomes more and more important. Thankfully, you have this study guide to help you get ready. Use this guide to help keep your studying on track and refer to it often.

This study guide contains several key sections that will help you be successful on your exam. The guide contains tips for what you should do the night before and the day of the test. Also included are test-taking tips. Knowing the right information is not always enough. Many well-prepared test takers struggle with exams. These tips will help equip you to accurately read, assess, and answer test questions.

A large part of the guide is devoted to showing you what content to expect on the exam and to helping you better understand that content. In this guide are practice test questions so that you can see how well you have grasped the content. Then, answer explanations are provided so that you can understand why you missed certain questions.

Don't try to cram the night before you take your exam. This is not a wise strategy for a few reasons. First, your retention of the information will be low. Your time would be better used by reviewing information you already know rather than trying to learn a lot of new information. Second, you will likely become stressed as you try to gain a large amount of knowledge in a short amount of time. Third, you will be depriving yourself of sleep. So be sure to go to bed at a reasonable time the night before. Being well-rested helps you focus and remain calm.

Be sure to eat a substantial breakfast the morning of the exam. If you are taking the exam in the afternoon, be sure to have a good lunch as well. Being hungry is distracting and can make it difficult to focus. You have hopefully spent lots of time preparing for the exam. Don't let an empty stomach get in the way of success!

When travelling to the testing center, leave earlier than needed. That way, you have a buffer in case you experience any delays. This will help you remain calm and will keep you from missing your appointment time at the testing center.

Be sure to pace yourself during the exam. Don't try to rush through the exam. There is no need to risk performing poorly on the exam just so you can leave the testing center early. Allow yourself to use all of the allotted time if needed.

Remain positive while taking the exam even if you feel like you are performing poorly. Thinking about the content you should have mastered will not help you perform better on the exam.

Once the exam is complete, take some time to relax. Even if you feel that you need to take the exam again, you will be well served by some down time before you begin studying again. It's often easier to convince yourself to study if you know that it will come with a reward!

Test-Taking Strategies

1. Predicting the Answer

When you feel confident in your preparation for a multiple-choice test, try predicting the answer before reading the answer choices. This is especially useful on questions that test objective factual knowledge. By predicting the answer before reading the available choices, you eliminate the possibility that you will be distracted or led astray by an incorrect answer choice. You will feel more confident in your selection if you read the question, predict the answer, and then find your prediction among the answer choices. After using this strategy, be sure to still read all of the answer choices carefully and completely. If you feel unprepared, you should not attempt to predict the answers. This would be a waste of time and an opportunity for your mind to wander in the wrong direction.

2. Reading the Whole Question

Too often, test takers scan a multiple-choice question, recognize a few familiar words, and immediately jump to the answer choices. Test authors are aware of this common impatience, and they will sometimes prey upon it. For instance, a test author might subtly turn the question into a negative, or he or she might redirect the focus of the question right at the end. The only way to avoid falling into these traps is to read the entirety of the question carefully before reading the answer choices.

3. Looking for Wrong Answers

Long and complicated multiple-choice questions can be intimidating. One way to simplify a difficult multiple-choice question is to eliminate all of the answer choices that are clearly wrong. In most sets of answers, there will be at least one selection that can be dismissed right away. If the test is administered on paper, the test taker could draw a line through it to indicate that it may be ignored; otherwise, the test taker will have to perform this operation mentally or on scratch paper. In either case, once the obviously incorrect answers have been eliminated, the remaining choices may be considered. Sometimes identifying the clearly wrong answers will give the test taker some information about the correct answer. For instance, if one of the remaining answer choices is a direct opposite of one of the eliminated answer choices, it may well be the correct answer. The opposite of obviously wrong is obviously right! Of course, this is not always the case. Some answers are obviously incorrect simply because they are irrelevant to the question being asked. Still, identifying and eliminating some incorrect answer choices is a good way to simplify a multiple-choice question.

4. Don't Overanalyze

Anxious test takers often overanalyze questions. When you are nervous, your brain will often run wild, causing you to make associations and discover clues that don't actually exist. If you feel that this may be a problem for you, do whatever you can to slow down during the test. Try taking a deep breath or counting to ten. As you read and consider the question, restrict yourself to the particular words used by the author. Avoid thought tangents about what the author *really* meant, or what he or she was *trying* to say. The only things that matter on a multiple-choice test are the words that are actually in the question. You must avoid reading too much into a multiple-choice question, or supposing that the writer meant something other than what he or she wrote.

5. No Need for Panic

It is wise to learn as many strategies as possible before taking a multiple-choice test, but it is likely that you will come across a few questions for which you simply don't know the answer. In this situation, avoid panicking. Because most multiple-choice tests include dozens of questions, the relative value of a single wrong answer is small. As much as possible, you should compartmentalize each question on a multiple-choice test. In other words, you should not allow your feelings about one question to affect your success on the others. When you find a question that you either don't understand or don't know how to answer, just take a deep breath and do your best. Read the entire question slowly and carefully. Try rephrasing the question a couple of different ways. Then, read all of the answer choices carefully. After eliminating obviously wrong answers, make a selection and move on to the next question.

6. Confusing Answer Choices

When working on a difficult multiple-choice question, there may be a tendency to focus on the answer choices that are the easiest to understand. Many people, whether consciously or not, gravitate to the answer choices that require the least concentration, knowledge, and memory. This is a mistake. When you come across an answer choice that is confusing, you should give it extra attention. A question might be confusing because you do not know the subject matter to which it refers. If this is the case, don't eliminate the answer before you have affirmatively settled on another. When you come across an answer choice of this type, set it aside as you look at the remaining choices. If you can confidently assert that one of the other choices is correct, you can leave the confusing answer aside. Otherwise, you will need to take a moment to try to better understand the confusing answer choice. Rephrasing is one way to tease out the sense of a confusing answer choice.

7. Your First Instinct

Many people struggle with multiple-choice tests because they overthink the questions. If you have studied sufficiently for the test, you should be prepared to trust your first instinct once you have carefully and completely read the question and all of the answer choices. There is a great deal of research suggesting that the mind can come to the correct conclusion very quickly once it has obtained all of the relevant information. At times, it may seem to you as if your intuition is working faster even than your reasoning mind. This may in fact be true. The knowledge you obtain while studying may be retrieved from your subconscious before you have a chance to work out the associations that support it. Verify your instinct by working out the reasons that it should be trusted.

8. Key Words

Many test takers struggle with multiple-choice questions because they have poor reading comprehension skills. Quickly reading and understanding a multiple-choice question requires a mixture of skill and experience. To help with this, try jotting down a few key words and phrases on a piece of scrap paper. Doing this concentrates the process of reading and forces the mind to weigh the relative importance of the question's parts. In selecting words and phrases to write down, the test taker thinks about the question more deeply and carefully. This is especially true for multiple-choice questions that are preceded by a long prompt.

9. Subtle Negatives

One of the oldest tricks in the multiple-choice test writer's book is to subtly reverse the meaning of a question with a word like *not* or *except*. If you are not paying attention to each word in the question, you can easily be led astray by this trick. For instance, a common question format is, "Which of the following is...?" Obviously, if the question instead is, "Which of the following is not...?," then the answer will be quite different. Even worse, the test makers are aware of the potential for this mistake and will include one answer choice that would be correct if the question were not negated or reversed. A test taker who misses the reversal will find what he or she believes to be a correct answer and will be so confident that he or she will fail to reread the question and discover the original error. The only way to avoid this is to practice a wide variety of multiple-choice questions and to pay close attention to each and every word.

10. Reading Every Answer Choice

It may seem obvious, but you should always read every one of the answer choices! Too many test takers fall into the habit of scanning the question and assuming that they understand the question because they recognize a few key words. From there, they pick the first answer choice that answers the question they believe they have read. Test takers who read all of the answer choices might discover that one of the latter answer choices is actually *more* correct. Moreover, reading all of the answer choices can remind you of facts related to the question that can help you arrive at the correct answer. Sometimes, a misstatement or incorrect detail in one of the latter answer choices will trigger your memory of the subject and will enable you to find the right answer. Failing to read all of the answer choices is like not reading all of the items on a restaurant menu: you might miss out on the perfect choice.

11. Spot the Hedges

One of the keys to success on multiple-choice tests is paying close attention to every word. This is never truer than with words like almost, most, some, and sometimes. These words are called "hedges" because they indicate that a statement is not totally true or not true in every place and time. An absolute statement will contain no hedges, but in many subjects, the answers are not always straightforward or absolute. There are always exceptions to the rules in these subjects. For this reason, you should favor those multiple-choice questions that contain hedging language. The presence of qualifying words indicates that the author is taking special care with their words, which is certainly important when composing the right answer. After all, there are many ways to be wrong, but there is only one way to be right! For this reason, it is wise to avoid answers that are absolute when taking a multiple-choice test. An absolute answer is one that says things are either all one way or all another. They often include words like *every*, *always*, *best*, and *never*. If you are taking a multiple-choice test in a subject that doesn't lend itself to absolute answers, be on your guard if you see any of these words.

12. Long Answers

In many subject areas, the answers are not simple. As already mentioned, the right answer often requires hedges. Another common feature of the answers to a complex or subjective question are qualifying clauses, which are groups of words that subtly modify the meaning of the sentence. If the question or answer choice describes a rule to which there are exceptions or the subject matter is complicated, ambiguous, or confusing, the correct answer will require many words in order to be expressed clearly and accurately. In essence, you should not be deterred by answer choices that seem excessively long. Oftentimes, the author of the text will not be able to write the correct answer without offering some qualifications and modifications. Your job is to read the answer choices thoroughly and completely and to select the one that most accurately and precisely answers the question.

13. Restating to Understand

Sometimes, a question on a multiple-choice test is difficult not because of what it asks but because of how it is written. If this is the case, restate the question or answer choice in different words. This process serves a couple of important purposes. First, it forces you to concentrate on the core of the question. In order to rephrase the question accurately, you have to understand it well. Rephrasing the question will concentrate your mind on the key words and ideas. Second, it will present the information to your mind in a fresh way. This process may trigger your memory and render some useful scrap of information picked up while studying.

14. True Statements

Sometimes an answer choice will be true in itself, but it does not answer the question. This is one of the main reasons why it is essential to read the question carefully and completely before proceeding to the answer choices. Too often, test takers skip ahead to the answer choices and look for true statements. Having found one of these, they are content to select it without reference to the question above. Obviously, this provides an easy way for test makers to play tricks. The savvy test taker will always read the entire question before turning to the answer choices. Then, having settled on a correct answer choice, he or she will refer to the original question and ensure that the selected answer is relevant. The mistake of choosing a correct-but-irrelevant answer choice is especially common on questions related to specific pieces of objective knowledge. A prepared test taker will have a wealth of factual knowledge at their disposal and should not be careless in its application.

15. No Patterns

One of the more dangerous ideas that circulates about multiple-choice tests is that the correct answers tend to fall into patterns. These erroneous ideas range from a belief that B and C are the most common right answers, to the idea that an unprepared test-taker should answer "A-B-A-C-A-D-A-B-A." It cannot be emphasized enough that pattern-seeking of this type is exactly the WRONG way to approach a multiple-choice test. To begin with, it is highly unlikely that the test maker will plot the correct answers according to some predetermined pattern. The questions are scrambled and delivered in a random order. Furthermore, even if the test maker was following a pattern in the assignation of correct answers, there is no reason why the test taker would know which pattern he or she was using. Any attempt to discern a pattern in the answer choices is a waste of time and a distraction from the real work of taking the test. A test taker would be much better served by extra preparation before the test than by reliance on a pattern in the answers.

FREE DVD OFFER

Don't forget that doing well on your exam includes both understanding the test content and understanding how to use what you know to do well on the test. We offer a completely FREE Test Taking Tips DVD that covers world class test taking tips that you can use to be even more successful when you are taking your test.

All that we ask is that you email us your feedback about your study guide. To get your **FREE Test Taking Tips DVD**, email freedvd@studyguideteam.com with "FREE DVD" in the subject line and the following information in the body of the email:

- The title of your study guide.
- Your product rating on a scale of 1-5, with 5 being the highest rating.
- Your feedback about the study guide. What did you think of it?
- Your full name and shipping address to send your free DVD.

Introduction to the GMAT

Function of the Test

The Graduate Management Admission Test (GMAT), administered by the Graduate Management Admissions Council (GMAC), is the standard business school admissions exam in the United States. More than 2,100 business schools use the GMAT as part of their admissions criteria for MBA programs as well as other business-related programs like accounting and finance. The test is also used in admissions offices at some business schools around the world. The GMAT is not the only business school admissions test—an increasing number of schools accept GRE scores also or instead—but it remains the most common in the United States.

Approximately 300,000 people take the GMAT every year, although that total varies with the economy and other factors. Although business schools admissions' departments do consider a variety of factors in making decisions about prospective students, the GMAT score is typically the primary consideration and the most commonly discussed factor.

Test Administration

The GMAT is offered at approximately 600 testing centers in over 100 countries. It is offered year-round, and prospective test takers should either register for the GMAT at mba.com or make an appointment at a time that the test is offered at their preferred location. Upon completion of the exam, a test taker may view their preliminary score on the multiple-choice sections. The test taker is then given two minutes in which to decide whether to report or cancel their score.

The multiple-choice sections of the GMAT are computer adaptive tests (CATs), meaning that they are given by computer and that the difficulty of questions offered to the test taker adapts to the test taker's performance up to that point. The essay portion is also conducted by computer, but scored by human graders.

Individuals are allowed to retake the GMAT as often as they wish, but not within thirty-one days of the last time they took the test (even if they cancelled their scores), and not more than five times in any one twelve-month span. Reasonable accommodations are available for test takers with disabilities in keeping with the Americans With Disabilities Act.

Test Format

The GMAT lasts about four hours, about thirty minutes of which consists of breaks. The verbal section is intended to measure the test-taker's ability to read, understand, and edit written English, and the ability to evaluate arguments. It contains forty-one questions divided among reading comprehension, sentence correction, and critical reasoning. There is little if any vocabulary tested on the GMAT.

The quantitative section covers arithmetic, algebra, and geometry. It typically contains both straight-forward problem-solving questions and also data sufficiency questions in which the test taker must determine whether the question provides enough information to accurately answer the question, and, if not, what additional information is required.

A summary of the sections of the GMAT is as follows:

Section	Questions	Time
Analytical Writing Assessment	1 question	30 minutes
Integrated Reasoning	12 questions	30 minutes
Quantitative Reasoning	31 questions	62 minutes
Verbal Reasoning	36 questions	65 minutes

Scoring

The GMAT score reported to schools is a scaled score ranging from 200 to 800 reflecting the test taker's combined performance on the verbal and quantitative sections of the exam. The analytical writing and integrated reasoning sections are scored separately, and are not typically given as much emphasis in schools' admission decisions. The average score of all test takers is usually around 540, and the standard deviation of all scores is 100 points, meaning that just over two-thirds of test takers fall in a range from 440 to 640. Top schools such as Harvard and Stanford admit students with average scores of around 725.

GMAT scores are based on the number of correct answers, with no penalty for guessing other than the missed opportunity to provide another correct response.

Recent/Future Developments

In July 2014, the GMAC began permitting test takers to view their preliminary scores before deciding whether to report or cancel them.

Study Prep Plan for the GMAT

1 **Schedule** - Use one of our study schedules below or come up with one of your own.

2 **Relax** - Test anxiety can hurt even the best students. There are many ways to reduce stress. Find the one that works best for you.

3 **Excecute** - Once you have a good plan in place, be sure to stick to it

Sample Study Plans

One Week Study Schedule

Day 1	Analytical Writing Assessment
Day 2	Integrated Reasoning
Day 3	Quantitative Reasoning
Day 4	Verbal Reasoning
Day 5	Practice Test #1
Day 6	Practice Test #2
Day 7	Take Your Exam!

Two Week Study Schedule

Day 1	Analyzing the Essay	Day 8	Sentence Correction
Day 2	Writing the Essay	Day 9	Practice Test #1
Day 3	Evaluating Information from Different Sources	Day 10	Review Answer Explanations #1
Day 4	Manipulating Information from Multiple Sources to Solve Problems	Day 11	Practice Test #2
Day 5	Data Sufficiency and Problem Solving	Day 12	Review Answer Explanations #2
Day 6	Reading Comprehension	Day 13	Review Questions from #1 and #2
Day 7	Critical Reasoning	Day 14	Take Your Exam!

One Month Study Schedule

Day 1	Analyzing the Essay	Day 11	Practice Test #2 – Integrated Reasoning	Day 21	Sentence Correction
Day 2	Writing the Essay	Day 12	Data Sufficiency and Problem Solving	Day 22	Practice Test #1 – Verbal Reasoning
Day 3	Conventions of Standard English	Day 13	Solving for X in Proportions	Day 23	Practice Test #2 – Verbal Reasoning
Day 4	AWA Prompt #1	Day 14	Translating Words into Math	Day 24	(Study Break)
Day 5	AWA Prompt #2	Day 15	Word Problems	Day 25	Review Answer Explanations #1 – Integrated Reasoning
Day 6	Graphics Interpretation	Day 16	Working with Money	Day 26	Review Answer Explanations #2 – Integrated Reasoning
Day 7	Evaluating Information from Different Sources	Day 17	Transformations of a Plane	Day 27	Review Answer Explanations #2 – Quantitative Reasoning
Day 8	Recognizing Relationships in the Information	Day 18	Practice Test #2 – Quantitative Reasoning	Day 28	Review Answer Explanations #1 – Verbal Reasoning
Day 9	Manipulating Information from Multiple Sources to Solve Problems	Day 19	Reading Comprehension	Day 29	Review Answer Explanations #2 – Verbal Reasoning
Day 10	Practice Test #1 – Integrated Reasoning	Day 20	Critical Reasoning	Day 30	Take Your Exam!

Analytical Writing Assessment

For the GMAT Analytical Writing Assessment, you will be given thirty minutes to read a given argument and write a response analyzing that same argument. Keep in mind that you are not being asked to present your own views on the topic; you are being asked to critique *how the argument is written*. Below are sections designed to help you analyze the given essay, write the response essay, and revise the response essay.

Analyzing the Essay

Understanding Organizational Patterns and Structures

Informational text is specifically designed to relate factual information, and although it is open to a reader's interpretation and application of the facts, the structure of the presentation is carefully designed to lead the reader to a particular conclusion or central idea. When reading informational text, it is important that readers are able to understand its organizational structure as the structure often directly relates to an author's intent to inform and/or persuade the reader.

The first step in identifying the text's structure is to determine the thesis or main idea. The thesis statement and organization of a work are closely intertwined. *A* **thesis statement** indicates the writer's purpose and may include the scope and direction of the text. It may be presented at the beginning of a text or at the end, and it may be explicit or implicit.

Once a reader has a grasp of the thesis or main idea of the text, he or she can better determine its organizational structure. Test takers are advised to read informational text passages more than once in order to comprehend the material fully. It is also helpful to examine any text features present in the text including the table of contents, index, glossary, headings, footnotes, and visuals. The analysis of these features and the information presented within them can offer additional clues about the central idea and structure of a text. The following questions should be asked when considering structure:

- How does the author assemble the parts to make an effective whole argument?
- Is the passage linear in nature and if so, what is the timeline or thread of logic?
- What is the presented order of events, facts, or arguments? Are these effective in contributing to the author's thesis?
- How can the passage be divided into sections? How are they related to each other and to the main idea or thesis?
- What key terms are used to indicate the organization?

Next, test takers should skim the passage, noting the first line or two of each body paragraph—the **topic sentences**—and the conclusion. Key **transitional terms**, such as *on the other hand*, *also*, *because*, *however*, *therefore*, *most importantly*, and *first*, within the text can also signal organizational structure. Based on these clues, readers should then be able to identify what type of organizational structure is being used.

The following organizational structures are most common:

- **Problem/solution**—organized by an analysis/overview of a problem, followed by potential solution(s)
- **Cause/effect**—organized by the effects resulting from a cause or the cause(s) of a particular effect
- **Spatial order**—organized by points that suggest location or direction—e.g., top to bottom, right to left, outside to inside
- **Chronological/sequence order**—organized by points presented to indicate a passage of time or through purposeful steps/stages
- **Comparison/Contrast**—organized by points that indicate similarities and/or differences between two things or concepts
- **Order of importance**—organized by priority of points, often most significant to least significant or vice versa

Understanding the Effect of Word Choice

An author's choice of words—also referred to as **diction**—helps to convey their meaning in a particular way. Through diction, an author can convey a particular tone—e.g., a humorous tone, a serious tone—in order to support the thesis in a meaningful way to the reader.

Connotation and Denotation

Connotation is when an author chooses words or phrases that invoke ideas or feelings other than their literal meaning. An example of the use of connotation is the word *cheap*, which suggests something is poor in value or negatively describes a person as reluctant to spend money. When something or someone is described this way, the reader is more inclined to have a particular image or feeling about it or him/her. Thus, connotation can be a very effective language tool in creating emotion and swaying opinion. However, connotations are sometimes hard to pin down because varying emotions can be associated with a word. Generally, though, connotative meanings tend to be fairly consistent within a specific cultural group.

Denotation refers to words or phrases that mean exactly what they say. It is helpful when a writer wants to present hard facts or vocabulary terms with which readers may be unfamiliar. Some examples of denotation are the words *inexpensive* and *frugal*. *Inexpensive* refers to the cost of something, not its value, and *frugal* indicates that a person is conscientiously watching their spending. These terms do not elicit the same emotions that *cheap* does.

Authors sometimes choose to use both, but what they choose and when they use it is what critical readers need to differentiate. One method isn't inherently better than the other; however, one may create a better effect, depending upon an author's intent. If, for example, an author's purpose is to inform, to instruct, and to familiarize readers with a difficult subject, their use of connotation may be helpful. However, it may also undermine credibility and confuse readers. An author who wants to create a credible, scholarly effect in their text would most likely use denotation, which emphasizes literal, factual meaning and examples.

Technical Language

Test takers and critical readers alike should be very aware of technical language used within informational text. **Technical language** refers to terminology that is specific to a particular industry and is best understood by those specializing in that industry. This language is fairly easy to differentiate, since it will most likely be unfamiliar to readers. It's critical to be able to define technical language either by the author's written definition, through the use of an included glossary—if offered—or through context clues that help readers clarify word meaning.

Rhetorical Strategies and Devices

A **rhetorical device** is the phrasing and presentation of an idea that reinforces and emphasizes a point in an argument. A rhetorical device is often quite memorable. One of the more famous uses of a rhetorical device is in John F. Kennedy's 1961 inaugural address: "Ask not what your country can do for you, ask what you can do for your country." The contrast of ideas presented in the phrasing is an example of the rhetorical device of antimetabole.

Some other common examples are provided below, but test takers should be aware that this is not a complete list.

Device	Definition	Example
Allusion	A reference to a famous person, event, or significant literary text as a form of significant comparison	"We are apt to shut our eyes against a painful truth, and listen to the song of that siren till she transforms us into beasts." Patrick Henry
Anaphora	The repetition of the same words at the beginning of successive words, phrases, or clauses, designed to emphasize an idea	"We shall not flag or fail. We shall go on to the end. We shall fight in France, we shall fight on the seas and oceans, we shall fight with growing confidence ... we shall fight in the fields and in the streets, we shall fight in the hills. We shall never surrender." Winston Churchill
Understatement	A statement meant to portray a situation as less important than it actually is to create an ironic effect	"The war in the Pacific has not necessarily developed in Japan's favor." Emperor Hirohito, surrendering Japan in World War II.
Parallelism	A syntactical similarity in a structure or series of structures used for impact of an idea, making it memorable	"A penny saved is a penny earned." Ben Franklin
Rhetorical question	A question posed that is not answered by the writer though there is a desired response, most often designed to emphasize a point	"Can anyone look at our reduced standing in the world today and say, 'Let's have four more years of this?'" Ronald Reagan

Rhetorical Appeals

In an argument or persuasive text, an author will strive to sway readers to an opinion or conclusion. To be effective, an author must consider their intended audience. Although an author may write text for a general audience, he or she will use methods of appeal or persuasion to convince that audience. Aristotle asserted that there were three methods or modes by which a person could be persuaded. These are referred to as **rhetorical appeals**.

The three main types of rhetorical appeals are shown in the following graphic.

Ethos, also referred to as an **ethical appeal**, is an appeal to the audience's perception of the writer as credible (or not), based on their examination of their ethics and who the writer is, their experience or incorporation of relevant information, or their argument. For example, authors may present testimonials to bolster their arguments. The reader who critically examines the veracity of the testimonials and the credibility of those giving the testimony will be able to determine if the author's use of testimony is valid to their argument. In turn, this will help the reader determine if the author's thesis is valid. An author's careful and appropriate use of technical language can create an overall knowledgeable effect and, in turn, act as a convincing vehicle when it comes to credibility. Overuse of technical language, however, may create confusion in readers and obscure an author's overall intent.

Pathos, also referred to as a **pathetic** or **emotional appeal**, is an appeal to the audience's sense of identity, self-interest, or emotions. A critical reader will notice when the author is appealing to pathos through anecdotes and descriptions that elicit an emotion such as anger or pity. Readers should also beware of factual information that uses generalization to appeal to the emotions. While it's tempting to believe an author is the source of truth in their text, an author who presents factual information as universally true, consistent throughout time, and common to all groups is using **generalization**. Authors who exclusively use generalizations without specific facts and credible sourcing are attempting to sway readers solely through emotion.

Logos, also referred to as a **logical appeal**, is an appeal to the audience's ability to see and understand the logic in a claim offered by the writer. A critical reader has to be able to evaluate an author's arguments for validity of reasoning and for sufficiency when it comes to argument.

Writing the Essay

Brainstorming

One of the most important steps in writing an essay is prewriting. Before drafting an essay, it's helpful to think about the topic for a moment or two, in order to gain a more solid understanding of the task. Then, spending about five minutes jotting down the immediate ideas that could work for the essay is recommended. It is a way to get some words on the page and offer a reference for ideas when drafting. Scratch paper is provided for writers to use any prewriting techniques such as webbing, free writing, or listing. The goal is to get ideas out of the mind and onto the page.

Considering Opposing Viewpoints

In the planning stage, it's important to consider all aspects of the topic, including different viewpoints on the subject. There are more than two ways to look at a topic, and a strong argument considers those opposing viewpoints. Considering opposing viewpoints can help writers present a fair, balanced, and informed essay that shows consideration for all readers. This approach can also strengthen an argument by recognizing and potentially refuting opposing viewpoint(s).

Moving from Brainstorming to Planning

Once the ideas are on the page, it's time to turn them into a solid plan for the essay. The best ideas from the brainstorming results can then be developed into a more formal outline. An outline typically has one main point (the thesis) and at least three sub-points that support the main point. Here's an example:

Main Idea

- Point #1
- Point #2
- Point #3

Of course, there will be details under each point, but this approach is the best for dealing with timed writing.

Staying on Track

Basing the essay on the outline aids in both organization and coherence. The goal is to ensure that there is enough time to develop each sub-point in the essay, roughly spending an equal amount of time on each idea. Keeping an eye on the time will help. If there are fifteen minutes left to draft the essay, then it makes sense to spend about 5 minutes on each of the ideas. Staying on task is critical to success, and timing out the parts of the essay can help writers avoid feeling overwhelmed.

Parts of the Essay

The **introduction** has to do a few important things:

- Establish the **topic** of the essay in original wording (i.e., not just repeating the prompt)
- Clarify the significance/importance of the topic or purpose for writing (not too many details, a brief overview)
- Offer a **thesis statement** that offers a critique of the prescribed conclusion (typically one-two brief sentences as a clear, concise explanation)

Body paragraphs reflect the ideas developed in the outline. Three-four points is probably sufficient for a short essay, and they should include the following:

- A **topic sentence** that identifies the sub-point (e.g., a reason why, a way how, a cause or effect)
- A detailed **explanation** of the point, explaining why the writer thinks this point is valid
- Illustrative **examples**, such as personal examples or real-world examples, that support and validate the point (i.e., "prove" the point)
- A **concluding sentence** that connects the examples, reasoning, and analysis to the point being made

The **conclusion**, or final paragraph, should be brief and should reiterate the focus, clarifying why the discussion is significant or important. It is important to avoid adding specific details or new ideas to this paragraph. The purpose of the conclusion is to sum up what has been said to bring the discussion to a close.

Don't Panic!

Writing an essay can be overwhelming, and performance panic is a natural response. The outline serves as a basis for the writing and helps to keep writers focused. Getting stuck can also happen, and it's helpful to remember that brainstorming can be done at any time during the writing process. Following the steps of the writing process is the best defense against writer's block.

Timed essays can be particularly stressful, but assessors are trained to recognize the necessary planning and thinking for these timed efforts. Using the plan above and sticking to it helps with time management. Timing each part of the process helps writers stay on track. Sometimes writers try to cover too much in their essays. If time seems to be running out, this is an opportunity to determine whether all of the ideas in the outline are necessary. Three body paragraphs are sufficient, and more than that is probably too much to cover in a short essay.

More isn't always *better* in writing. A strong essay will be clear and concise. It will avoid unnecessary or repetitive details. It is better to have a concise, five-paragraph essay that makes a clear point, than a ten-paragraph essay that doesn't. The goal is to write one-two pages of quality writing. Paragraphs should also reflect balance; if the introduction goes to the bottom of the first page, the writing may be going off-track or be repetitive. It's best to fall into the one-two page range, but a complete, well-developed essay is the ultimate goal.

The Final Steps

Leaving a few minutes at the end to revise and proofread offers an opportunity for writers to polish things up. Putting one's self in the reader's shoes and focusing on what the essay actually says helps writers identify problems—it's a movement from the mindset of writer to the mindset of editor. The goal is to have a clean, clear copy of the essay. The following areas should be considered when proofreading:

- Sentence fragments
- Awkward sentence structure
- Run-on sentences
- Incorrect word choice
- Grammatical agreement errors
- Spelling errors
- Punctuation errors
- Capitalization errors

The Short Overview

The essay may seem challenging, but following these steps can help writers focus:

- Take one-two minutes to think about the topic.
- Generate some ideas through brainstorming (three-four minutes).
- Organize ideas into a brief outline, selecting just three-four main points to cover in the essay (eventually the body paragraphs).
- Develop essay in parts:
- Introduction paragraph, with intro to topic and main points
- Viewpoint on the subject at the end of the introduction
- Body paragraphs, based on outline
- Each paragraph: makes a main point, explains the viewpoint, uses examples to support the point
- Brief conclusion highlighting the main points and closing
- Read over the essay (last five minutes).
- Look for any obvious errors, making sure that the writing makes sense.

Conventions of Standard English

Sentence Fragments

A **complete sentence** requires a verb and a subject that expresses a complete thought. Sometimes, the subject is omitted in the case of the implied *you*, used in sentences that are the command or imperative form—e.g., "Look!" or "Give me that." It is understood that the subject of the command is *you*, the listener or reader, so it is possible to have a structure without an explicit subject. Without these elements, though, the sentence is incomplete—it is a **sentence fragment**. While sentence fragments often occur in conversational English or creative writing, they are generally not appropriate in academic writing. Sentence fragments often occur when dependent clauses are not joined to an independent clause:

> **Sentence fragment**: Because the airline overbooked the flight.

The sentence above is a dependent clause that does not express a complete thought. What happened as a result of this cause? With the addition of an independent clause, this now becomes a complete sentence:

> **Complete sentence**: Because the airline overbooked the flight, several passengers were unable to board.

Sentences fragments may also occur through improper use of conjunctions:

> I'm going to the Bahamas for spring break. And to New York City for New Year's Eve.

While the first sentence above is a complete sentence, the second one is not because it is a prepositional phrase that lacks a subject [I] and a verb [am going]. Joining the two together with the coordinating conjunction forms one grammatically-correct sentence:

> I'm going to the Bahamas for spring break and to New York City for New Year's Eve.

Run-Ons

A **run-on** is a sentence with too many independent clauses that are improperly connected to each other:

> This winter has been very cold some farmers have suffered damage to their crops.

The sentence above has two subject-verb combinations. The first is "this winter has been"; the second is "some farmers have suffered." However, they are simply stuck next to each other without any punctuation or conjunction. Therefore, the sentence is a run-on.

Another type of run-on occurs when writers use inappropriate punctuation:

> This winter has been very cold, some farmers have suffered damage to their crops.

Though a comma has been added, this sentence is still not correct. When a comma alone is used to join two independent clauses, it is known as a **comma splice**. Without an appropriate conjunction, a comma cannot join two independent clauses by itself.

Run-on sentences can be corrected by either dividing the independent clauses into two or more separate sentences or inserting appropriate conjunctions and/or punctuation. The run-on sentence can be amended by separating each subject-verb pair into its own sentence:

> This winter has been very cold. Some farmers have suffered damage to their crops.

The run-on can also be fixed by adding a comma and conjunction to join the two independent clauses with each other:

> This winter has been very cold, so some farmers have suffered damage to their crops.

Parallelism

Parallel structure occurs when phrases or clauses within a sentence contain the same structure. Parallelism increases readability and comprehensibility because it is easy to tell which sentence elements are paired with each other in meaning.

> Jennifer enjoys cooking, knitting, and to spend time with her cat.

This sentence is not parallel because the items in the list appear in two different forms. Some are **gerunds**, which is the verb + ing: *cooking, knitting*. The other item uses the **infinitive** form, which is to + verb: *to spend*. To create parallelism, all items in the list may reflect the same form:

> Jennifer enjoys cooking, knitting, and spending time with her cat.

All of the items in the list are now in gerund forms, so this sentence exhibits parallel structure. Here's another example:

> The company is looking for employees who are responsible and with a lot of experience.

Again, the items that are listed in this sentence are not parallel. "Responsible" is an adjective, yet "with a lot of experience" is a prepositional phrase. The sentence elements do not utilize parallel parts of speech.

> The company is looking for employees who are responsible and experienced.

"Responsible" and "experienced" are both adjectives, so this sentence now has parallel structure.

Dangling and Misplaced Modifiers

Modifiers enhance meaning by clarifying or giving greater detail about another part of a sentence. However, incorrectly-placed modifiers have the opposite effect and can cause confusion. A **misplaced modifier** is a modifier that is not located appropriately in relation to the word or phrase that it modifies:

> Because he was one of the greatest thinkers of Renaissance Italy, John idolized Leonardo da Vinci.

In this sentence, the modifier is "because he was one of the greatest thinkers of Renaissance Italy," and the noun it is intended to modify is "Leonardo da Vinci." However, due to the placement of the modifier next to the subject, John, it seems as if the sentence is stating that John was a Renaissance genius, not Da Vinci.

> John idolized Leonard da Vinci because he was one of the greatest thinkers of Renaissance Italy.

The modifier is now adjacent to the appropriate noun, clarifying which of the two men in this sentence is the greatest thinker.

Dangling modifiers modify a word or phrase that is not readily apparent in the sentence. That is, they "dangle" because they are not clearly attached to anything:

> After getting accepted to college, Amir's parents were proud.

The modifier here, "after getting accepted to college," should modify who got accepted. The noun immediately following the modifier is "Amir's parents"—but they are probably not the ones who are going to college.

> After getting accepted to college, Amir made his parents proud.

The subject of the sentence has been changed to Amir himself, and now the subject and its modifier are appropriately matched.

Inconsistent Verb Tense

Verb tense reflects when an action occurred or a state existed. For example, the tense known as **simple present** expresses something that is happening right now or that happens regularly:

> She *works* in a hospital.

Present continuous tense expresses something in progress. It is formed by to be + verb + -ing.

> Sorry, I can't go out right now. I *am doing* my homework.

Past tense is used to describe events that previously occurred. However, in conversational English, speakers often use present tense or a mix of past and present tense when relating past events because it gives the narrative a sense of immediacy. In formal written English, though, consistency in verb tense is necessary to avoid reader confusion.

> I traveled to Europe last summer. As soon as I stepped off the plane, I feel like I'm in a movie! I'm surrounded by quaint cafes and impressive architecture.

The passage above abruptly switches from past tense—*traveled, stepped*—to present tense—*feel, am surrounded*.

> I *traveled* to Europe last summer. As soon as I *stepped* off the plane, I *felt* like I was in a movie! I *was surrounded* by quaint cafes and impressive architecture.

All verbs are in past tense, so this passage now has consistent verb tense.

Split Infinitives

The **infinitive form** of a verb consists of "to + base verb"—e.g., to walk, to sleep, to approve. A **split infinitive** occurs when another word, usually an adverb, is placed between *to* and the verb:

> I decided *to simply walk* to work to get more exercise every day.

The infinitive *to walk* is split by the adverb *simply*.

> It was a mistake *to hastily approve* the project before conducting further preliminary research.

The infinitive *to approve* is split by *hastily*.

Although some grammarians still advise against split infinitives, this syntactic structure is common in both spoken and written English and is widely accepted in standard usage.

Subject-Verb Agreement

In English, verbs must agree with the subject. The form of a verb may change depending on whether the subject is singular or plural, or whether it is first, second, or third person. For example, the verb *to be* has various forms:

I am a student.

You are a student.

She is a student.

We are students.

They are students.

Errors occur when a verb does not agree with its subject. Sometimes, the error is readily apparent:

We is hungry.

Is is not the appropriate form of *to be* when used with the third person plural *we*.

We are hungry.

This sentence now has correct subject-verb agreement.

However, some cases are trickier, particularly when the subject consists of a lengthy noun phrase with many modifiers:

Students who are hoping to accompany the anthropology department on its annual summer trip to Ecuador needs to sign up by March 31st.

The verb in this sentence is *needs*. However, its subject is not the noun adjacent to it—Ecuador. The subject is the noun at the beginning of the sentence—students. Because *students* is plural, *needs* is the incorrect verb form.

Students who are hoping to accompany the anthropology department on its annual summer trip to Ecuador *need* to sign up by March 31st.

This sentence now uses correct agreement between *students* and *need*.

Another case to be aware of is a **collective noun**. A collective noun refers to a group of many things or people but can be singular in itself—e.g., family, committee, army, pair team, council, jury. Whether or not a collective noun uses a singular or plural verb depends on how the noun is being used. If the noun refers to the group performing a collective action as one unit, it should use a singular verb conjugation:

The family is moving to a new neighborhood.

The whole family is moving together in unison, so the singular verb form *is* is appropriate here.

The committee has made its decision.

The verb *has* and the possessive pronoun *its* both reflect the word *committee* as a singular noun in the sentence above; however, when a collective noun refers to the group as individuals, it can take a plural verb:

The newlywed pair spend every moment together.

This sentence emphasizes the love between two people in a pair, so it can use the plural verb *spend*.

The council are all newly elected members.

The sentence refers to the council in terms of its individual members and uses the plural verb *are*.

Overall, though, American English is more likely to pair a collective noun with a singular verb, while British English is more likely to pair a collective noun with a plural verb.

Integrated Reasoning

Graphics Interpretation

Identifying Information from a Graphic

Texts may have graphic representations to help illustrate and visually support assertions made. For example, graphics can be used to express samples or segments of a population or demonstrate growth or decay. Three of the most popular graphic formats include line graphs, bar graphs, and pie charts.

Line graphs rely on a horizontal X-axis and a vertical Y-axis to establish baseline values. A point is plotted for each data value where the x-value and y-value of the data point intersect the axes, and those points are connected with lines. Compared to bar graphs or pie charts, line graphs are more useful for looking at the past and present and predicting future outcomes. For instance, a potential investor would look for stocks that demonstrated steady growth over many decades when examining the stock market. Note that severe spikes up and down indicate instability, while line graphs that display a slow but steady increase may indicate good returns.

Here's an example of a bar graph:

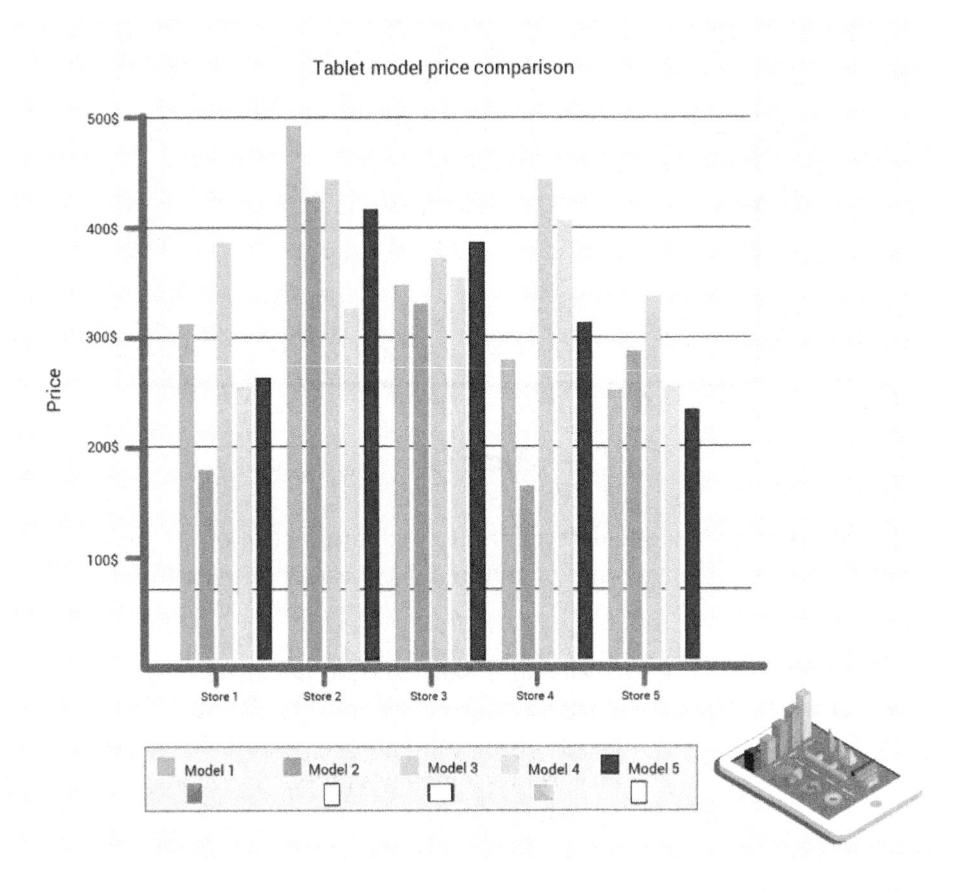

Bar graphs are usually displayed on a vertical Y-axis. The bars themselves can be two- or three-dimensional, depending on the designer's tastes. Unlike a line graph, which shows the fluctuation of only one variable, the X-axis on a bar graph is excellent for making comparisons because it shows differences between several variables. For instance, if an electronics store wanted to visually represent the number tablet sales for the year, a bar graph could have a bar for each type of tablet offered. To provide additional information, the store could show quarterly sales by constructing a bar for each type of tablet for each quarter in the fiscal year. The height of the bar would indicate the number of sales. The tablet types would be displayed along the x-axis with groups of four bars per tablet—one for each quarter.

Evaluating Information from Different Sources

When approaching the integrated reasoning sections of the GMAT, it is important to use all the information provided. Such information can be given from multiple sources, including tables, charts, graphs, and small excerpts from emails or articles. When viewing information, separate out key pieces, such as specific numbers that might determine a trend within a table or in a report of performance or sales. All sources should be labeled, thus making it easier to ascertain the relevance or importance of the data. Take an overall view of a table for the range of numbers and the headings of each column and row. On graphs or charts, the *x* and *y* axes (or horizontal and vertical lines) should be labeled with the value they are measuring, along with the base units for the measurements.

For example, the following position versus time graph is measured in meters per second, and the slope of the line can be determined to find the velocity.

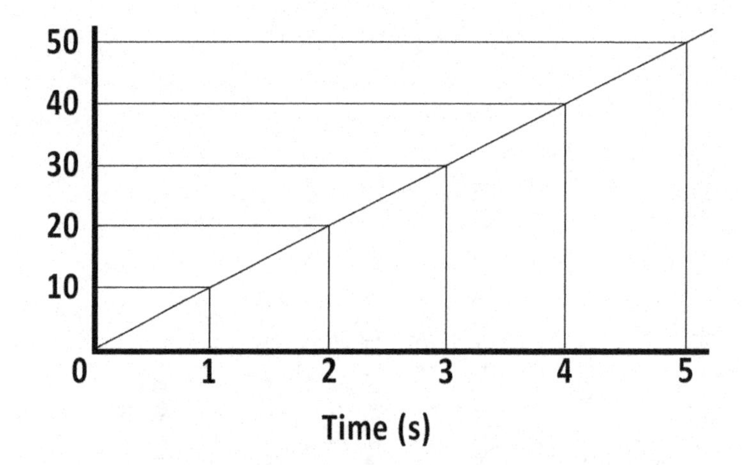

These pieces of information are used to determine past and future positions of the object being measured. A comparison of the movements of another object can determine trends in the positions of both objects.

The source of information from written dialogue, emails, or content can be used to determine the relevance of the material to the overall scope of the information. For example, an email from a company's marketing expert could contain more insight regarding a product's sales potential within a certain demographic compared to an email from a company's production manager. The source of the information determines the weight it should be given.

Another method for relaying information is tables or charts. These can be singular and compare multiple levels all in one place, like the following examples.

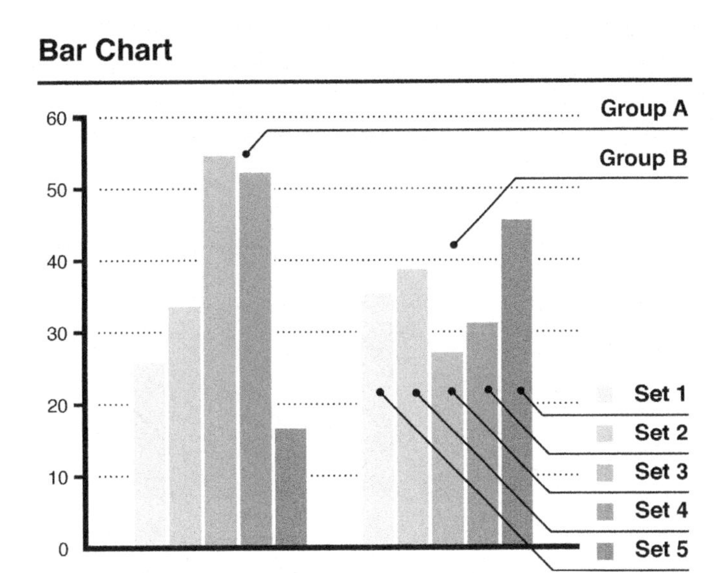

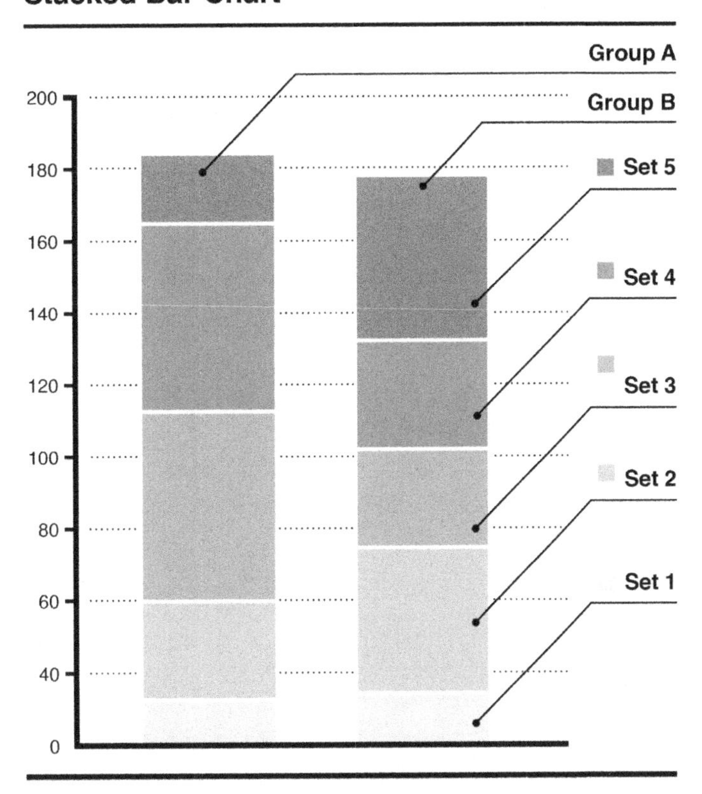

Again, it is important to identify what is being represented on each axis and to understand the relationship among the data being displayed. Sometimes, there is no relationship or correlation among the data points; it is simply a concise way to represent a larger amount of data.

An important item to take note of is the size of the units being represented. A misinterpretation of single units versus thousands of units can alter how information is read from the graph or compared from one data set to another. For example, the following two graphs display the same units (force versus time) but represent very different numbers when finding a point (impulse) on the graph. It may not seem like much of a difference; however, if a company was attempting to compare the results from an impact test, the difference between Newtons and kilonewtons is a factor of 1000. This could mean the difference between a dent in a bumper and a fatal crash. This type of oversight could cost lives, someone's employment, and possibly topple a business.

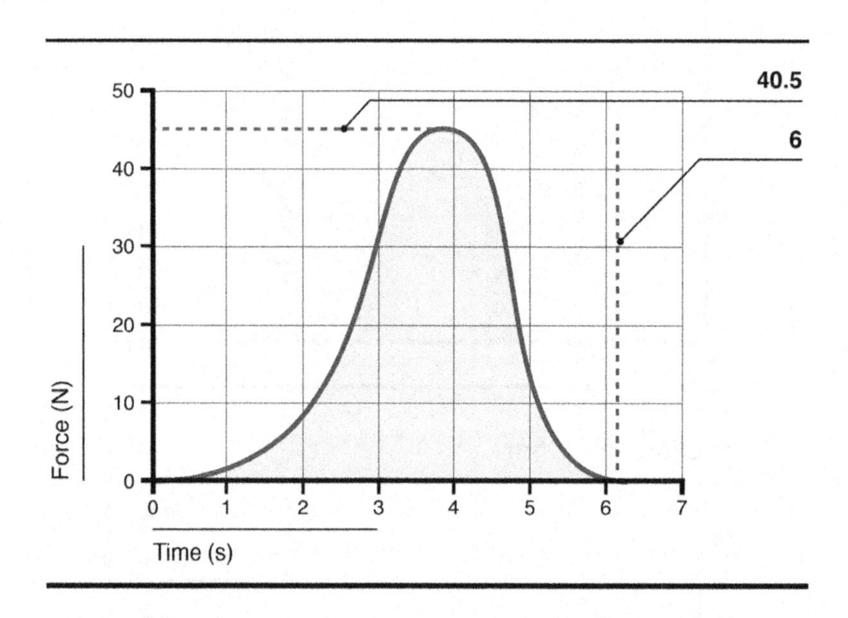

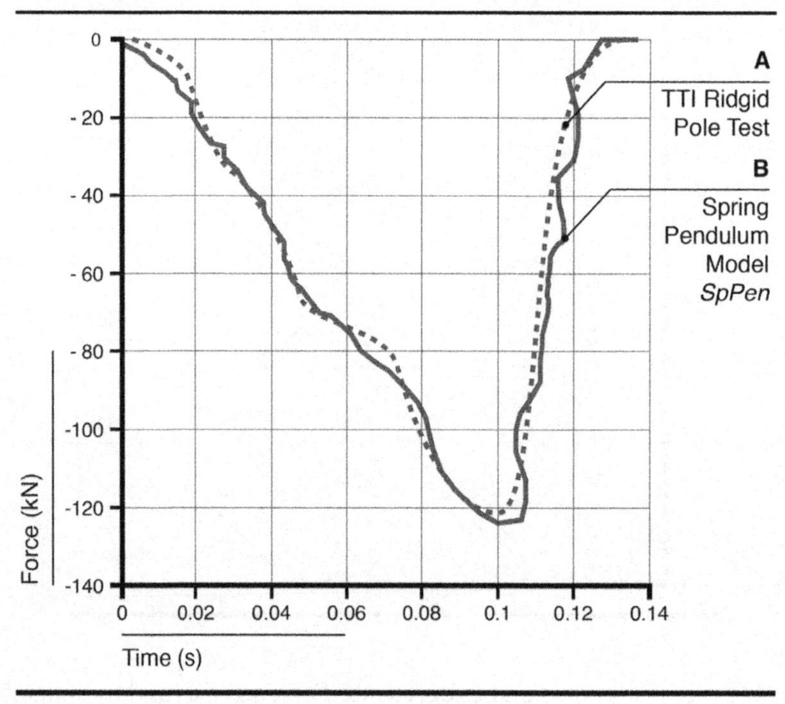

Recognizing Relationships in the Information

Information can also be provided by tables or graphs in a concise manner. A pie chart can depict multiple layers of information, especially when compared against other similarly constructed charts.

For example, the following pie charts can demonstrate sizable amounts of data over the course of time.

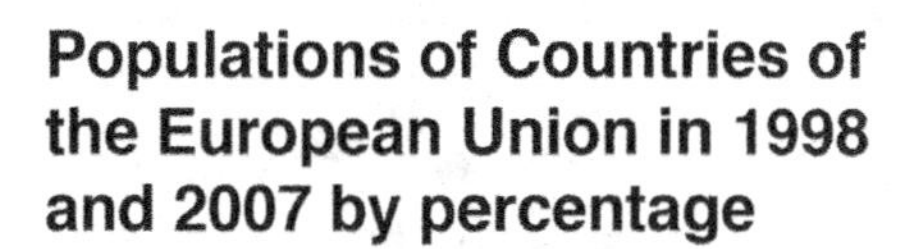

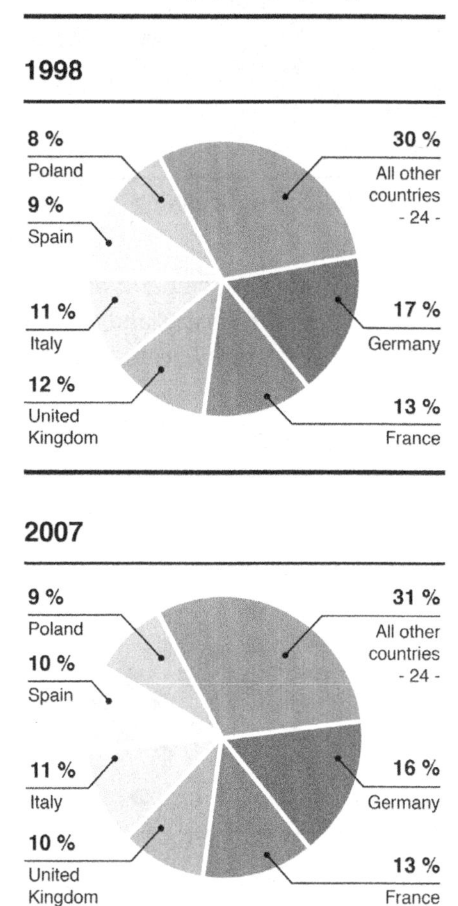

When putting such charts next to each other, it is easy to look at similarities and differences over the course of time, or among elements of the charts. The importance of such tools is observed in the ease of comparing multiple strings of data at once rather than being limited to simply one or two variables.

Relationships between similar measures can provide insight into correlations from year to year or product to product.

For example, the following graph can show a close positive correlation between the time students spend studying and their grades. This means that as the time spent studying increases, the students' grades also increase.

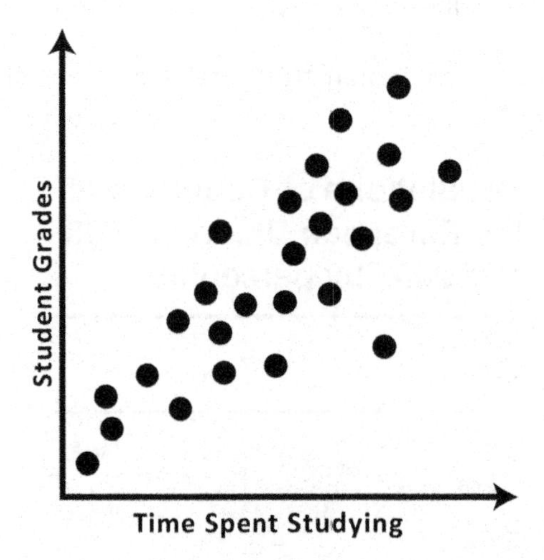

There can also be a negative correlation between variables. For example, the following graph shows a negative correlation between the number of missed classes and a student's exam score. It can be observed that the more classes a student misses, the lower the student's scores register.

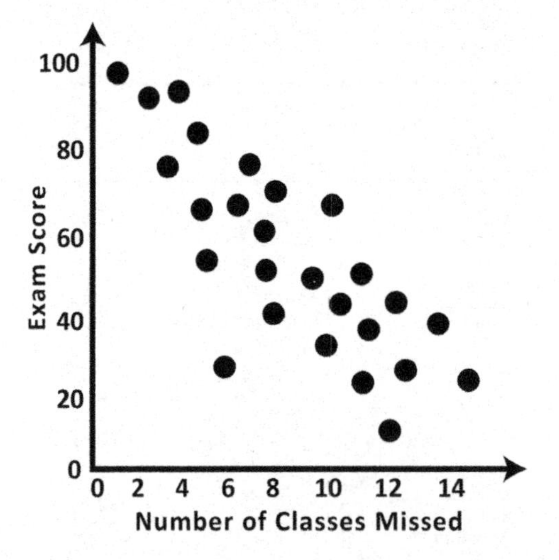

It can be a little less obvious when trying to identify relationships in information given via emails, or opinion-based writings. Generally, most business-based communications are dealing with a common topic that would need to be addressed at a meeting, such as a product, a marketing strategy, or production cost versus profit estimate. It can be difficult to decipher what the motivating factor might be for an employee's stance, but often the employee's role in the company can disclose what area he or she is going to regard as most important.

For example, a letter from the marketing expert in a company might discuss the best way to get a product out to the public and begin to register sales, while a financial expert for the same company may want to discuss the same product but discontinue its production due to the lack of profit potential compared to some unforeseen production costs. While both have the same topic in mind for discussion at a meeting, their stances on the future of the product differ. It is important to analyze any information given by each person's writing to understand their motivating factors and, thus, their position on a topic.

Manipulating Information from Multiple Sources to Solve Problems

Often information/data can be provided in a rawer format that needs a bit of manipulation to find the desired answer. An example of this would be a table of data collected from an experiment or a survey taken from a sample. This type of data will often need to be rearranged so that comparisons can be made among the represented items. For example, the following information could be sorted in several ways to extrapolate specific information. The table could be sorted (as displayed) to determine the peak months for T-shirt purchases, or it could be re-sorted to determine the most purchased T-shirt size. Both pieces of information would come from the same data set, both would be important for production estimates, and both would require some manipulation of the information given.

	D	E	F
1	**Payment**	**T-shirt Color**	**T-shirt Size**
2	1 - Jan	**Heather Gray**	*Large*
3	1 - Jan	**White**	*Large*
4	4 - Jan	**Dark Red**	*X - Large*
5	5 - Jan	**Dark Red**	*Medium*
6	5 - Jan	**Heather Gray**	*Large*
7	5 - Jan	**Dark Red**	*Medium*
8	5 - Jan	**Heather Gray**	*X - Large*
9	6 - Jan	**White**	*X - Large*
10	6 - Jan	**Dark Red**	*X - Large*
11	7 - Jan	**Heather Gray**	*Small*
12	7 - Jan	**Dark Red**	*Small*
13	7 - Jan	**Heather Gray**	*Small*
14	7 - Jan	**Heather Gray**	*Small*
15	11 - Jan	**Dark Red**	*Medium*
16	11 - Jan	**White**	*Medium*
17	11 - Jan	**Dark Red**	*Medium*

In some cases, multiple pieces of information from various sources need to be used to assess a situation, solve a problem, or determine a sound course of action. For example, the following set of sources could be used to determine the best way for a small business to cater to its focus audience for advertising, sharing information, or simply networking. Each piece of information would need to be cross-referenced from one set of tables to the other to match up the best social media source for the desired small business goal.

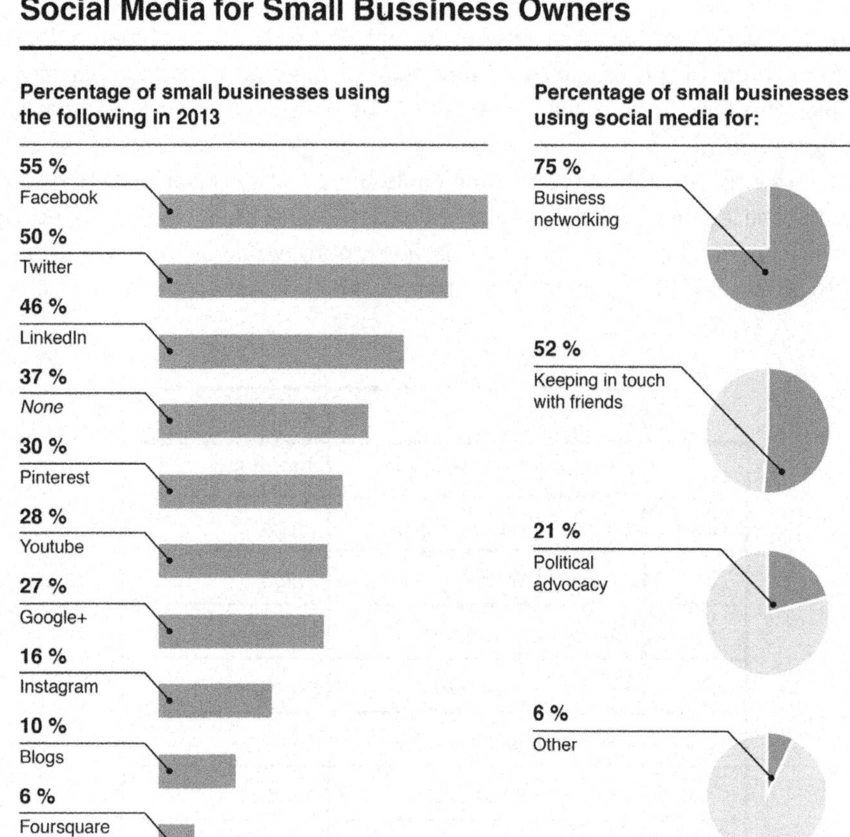

Gathering information from multiple sources and then reapplying or manipulating it to make comparisons or estimates is the essence of what propels businesses. The importance of this can be seen in the cohesive branches of a business needing to communicate key pieces of information so the other branches can make decisions to further the company's overall goal. For instance, a T-shirt company needs to cross-reference production costs, availability of production products (such as textiles and dyes), implementation costs, employee costs and benefits, marketing strategies, and potential sales during both low and peak times. Additional promotions used for increasing sales could be observed and mapped onto existing data tables and manipulated to determine if the expense of the trial promotion produced enough profit to repeat.

Oftentimes, using information from multiple sources can assist with solving a basic problem. To hypothesize if a new idea would be profitable, it is recommended that some research be done to see if there are any types of similar productions, promotions, marketing, etc. This would require conducting a type of literature review and extracting any pertinent information that could be applied and manipulated to make estimates or predictions toward the idea being developed.

This requires a great deal of comparisons across multiple venues. When looking to make such comparisons, existing data is not always in the exact form necessary to draw a solid conclusion. Therefore, the research much be augmented by any predicted standard deviations or alternative results that could arise. This should include being able to make some predictions for plans a business might have involving the idea. For example, if a toy company wanted to produce a new toy that involved wheels, they would need to research similar toys that use the same type/size of wheels and decide if there were age restrictions for users, cost restrictions for production, durability restrictions for the use of the toy, and the overall aesthetics involved with the existing toys that use that type of wheel. This kind of information manipulation could aid in cutting costs, focusing marketing, and avoiding safety issues.

Quantitative Reasoning

Data Sufficiency and Problem Solving

Arithmetic

Addition with Whole Numbers and Fractions

Addition combines two quantities together. With whole numbers, this is taking two sets of things and merging them into one, then counting the result. For example, $4 + 3 = 7$. When adding numbers, the order does not matter: $3 + 4 = 7$, also. Longer lists of whole numbers can also be added together. The result of adding numbers is called the **sum**.

With fractions, the number on top is the **numerator**, and the number on the bottom is the **denominator**. To add fractions, the denominator must be the same—a **common denominator**. To find a common denominator, the existing numbers on the bottom must be considered, and the lowest number they will both multiply into must be determined. Consider the following equation:

$$\frac{1}{3}+\frac{5}{6}=?$$

The numbers 3 and 6 both multiply into 6. Three can be multiplied by 2, and 6 can be multiplied by 1. The top and bottom of each fraction must be multiplied by the same number. Then, the numerators are added together to get a new numerator. The following equation is the result:

$$\frac{1}{3}+\frac{5}{6}=\frac{2}{6}+\frac{5}{6}=\frac{7}{6}$$

Subtraction with Whole Numbers and Fractions

Subtraction is taking one quantity away from another, so it is the opposite of addition. The expression $4 - 3$ means taking 3 away from 4. So, $4 - 3 = 1$. In this case, the order matters, since it entails taking one quantity away from the other, rather than just putting two quantities together. The result of subtraction is also called the **difference**.

To subtract fractions, the denominator must be the same. Then, subtract the numerators together to get a new numerator. Here is an example:

$$\frac{1}{3}-\frac{5}{6}=\frac{2}{6}-\frac{5}{6}=\frac{-3}{6}=-\frac{1}{2}$$

Multiplication with Whole Numbers and Fractions

Multiplication is a kind of repeated addition. The expression 4×5 is taking four sets, each of them having five things in them, and putting them all together. That means:

$$4 \times 5 = 5 + 5 + 5 + 5 = 20$$

As with addition, the order of the numbers does not matter. The result of a multiplication problem is called the **product**.

To multiply fractions, the numerators are multiplied to get the new numerator, and the denominators are multiplied to get the new denominator:

$$\frac{1}{3} \times \frac{5}{6} = \frac{1 \times 5}{3 \times 6} = \frac{5}{18}$$

When multiplying fractions, common factors can **cancel** or **divide into one another**, when factors that appear in the numerator of one fraction and the denominator of the other fraction. Here is an example:

$$\frac{1}{3} \times \frac{9}{8} = \frac{1}{1} \times \frac{3}{8}$$

$$1 \times \frac{3}{8} = \frac{3}{8}$$

The numbers 3 and 9 have a common factor of 3, so that factor can be divided out.

Division with Whole Numbers and Fractions

Division is the opposite of multiplication. With whole numbers, it means splitting up one number into sets of equal size. For example, $16 \div 8$ is the number of sets of eight things that can be made out of sixteen things. Thus, $16 \div 8 = 2$. As with subtraction, the order of the numbers will make a difference, here. The answer to a division problem is called the **quotient**, while the number in front of the division sign is called the **dividend,** and the number behind the division sign is called the **divisor**.

To divide fractions, the first fraction must be multiplied with the reciprocal of the second fraction. The **reciprocal** of the fraction $\frac{x}{y}$ is the fraction $\frac{y}{x}$. Here is an example:

$$\frac{1}{3} \div \frac{5}{6} = \frac{1}{3} \times \frac{6}{5} = \frac{6}{15} = \frac{2}{5}$$

Recognizing Equivalent Fractions and Mixed Numbers

The value of a fraction does not change if multiplying or dividing both the numerator and the denominator by the same number (other than 0). In other words, $\frac{x}{y} = \frac{a \times x}{a \times y} = \frac{x \div a}{y \div a}$, as long as *a* is not 0. This means that $\frac{2}{5} = \frac{4}{10}$, for example. If *x* and *y* are integers that have no common factors, then the fraction is said to be **simplified**. This means $\frac{2}{5}$ is simplified, but $\frac{4}{10}$ is not.

Often when working with fractions, the fractions need to be rewritten so that they all share a single denominator—this is called finding a **common denominator** for the fractions. Using two fractions, $\frac{a}{b}$ and $\frac{c}{d}$, the numerator and denominator of the left fraction can be multiplied by *d*, while the numerator and denominator of the right fraction can be multiplied by *b*. This provides the fractions $\frac{a \times d}{b \times d}$ and $\frac{c \times b}{d \times b}$ with the common denominator $b \times d$.

A fraction whose numerator is smaller than its denominator is called a **proper fraction**. A fraction whose numerator is bigger than its denominator is called an **improper fraction**. These numbers can be rewritten as a combination of integers and fractions, called a **mixed number**. For example:

$$\frac{6}{5} = \frac{5}{5} + \frac{1}{5} = 1 + \frac{1}{5} \text{ which can be written as: } 1\frac{1}{5}$$

Estimating

Estimation is finding a value that is close to a solution but is not the exact answer. For example, if there are values in the thousands to be multiplied, then each value can be estimated to the nearest thousand and the calculation performed. This value provides an approximate solution that can be determined very quickly.

Recognition of Decimals

The **decimal system** is a way of writing out numbers that uses ten different numerals: 0, 1, 2, 3, 4, 5, 6, 7, 8, and 9. This is also called a "base ten" or "base 10" system. Other bases are also used. For example, computers work with a base of 2. This means they only use the numerals 0 and 1.

The **decimal place** denotes how far to the right of the decimal point a numeral is. The first digit to the right of the decimal point is in the *tenths* place. The next is the *hundredths*. The third is the *thousandths*.

So, 3.142 has a 1 in the tenths place, a 4 in the hundredths place, and a 2 in the thousandths place.

The **decimal point** is a period used to separate the *ones* place from the *tenths* place when writing out a number as a decimal.

A **decimal number** is a number written out with a decimal point instead of as a fraction, for example, 1.25 instead of $\frac{5}{4}$. Depending on the situation, it can sometimes be easier to work with fractions and sometimes easier to work with decimal numbers.

A decimal number is **terminating** if it stops at some point. It is called **repeating** if it never stops but repeats a pattern over and over. It is important to note that every rational number can be written as a terminating decimal or as a repeating decimal.

Addition with Decimals

To add decimal numbers, each number in columns needs to be lined up by the decimal point. For each number being added, the zeros to the right of the last number need to be filled in so that each of the numbers has the same number of places to the right of the decimal. Then, the columns can be added together. Here is an example of $2.45 + 1.3 + 8.891$ written in column form:

$$\begin{array}{r} 2.450 \\ 1.300 \\ \underline{+\ 8.891} \end{array}$$

Zeros have been added in the columns so that each number has the same number of places to the right of the decimal.

Added together, the correct answer is 12.641:

$$\begin{array}{r} 2.450 \\ 1.300 \\ \underline{+\ 8.891} \\ 12.641 \end{array}$$

Subtraction with Decimals

Subtracting decimal numbers is the same process as adding decimals. Here is 7.89 – 4.235 written in column form:

$$\begin{array}{r} 7.890 \\ -\underline{4.235} \\ 3.655 \end{array}$$

A zero has been added in the column so that each number has the same number of places to the right of the decimal.

Multiplication with Decimals

The simplest way to multiply decimals is to calculate the product as if the decimals are not there, then count the number of decimal places in the original problem. Use that total to place the decimal the same number of places over in your answer, counting from right to left. For example, 0.5×1.25 can be rewritten and multiplied as 5×125, which equals 625. Then the decimal is added three places from the right for .625.

The final answer will have the same number of decimal points as the total number of decimal places in the problem. The first number has one decimal place, and the second number has two decimal places. Therefore, the final answer will contain three decimal places:

$$0.5 \times 1.25 = 0.625$$

Division with Decimals

Dividing a decimal by a whole number entails using long division first by ignoring the decimal point. Then, the decimal point is moved the number of places given in the problem.

For example, $6.8 \div 4$ can be rewritten as $68 \div 4$, which is 17. There is one non-zero integer to the right of the decimal point, so the final solution would have one decimal place to the right of the solution. In this case, the solution is 1.7.

Dividing a decimal by another decimal requires changing the divisor to a whole number by moving its decimal point. The decimal place of the dividend should be moved by the same number of places as the divisor. Then, the problem is the same as dividing a decimal by a whole number.

For example, $5.72 \div 1.1$ has a divisor with one decimal point in the denominator. The expression can be rewritten as $57.2 \div 11$ by moving each number one decimal place to the right to eliminate the decimal. The long division can be completed as $572 \div 11$ with a result of 52. Since there is one non-zero integer to the right of the decimal point in the problem, the final solution is 5.2.

In another example, $8 \div 0.16$ has a divisor with two decimal points in the denominator. The expression can be rewritten as $800 \div 16$ by moving each number two decimal places to the right to eliminate the decimal in the divisor. The long division can be completed with a result of 50.

Fraction and Percent Equivalencies

The word **percent** comes from the Latin phrase for "per one hundred." A percent is a way of writing out a fraction. It is a fraction with a denominator of 100. Thus, $65\% = \frac{65}{100}$.

To convert a fraction to a percent, the denominator is written as 100. For example:

$$\frac{3}{5} = \frac{60}{100} = 60\%$$

In converting a percent to a fraction, the percent is written with a denominator of 100, and the result is simplified. For example:

$$30\% = \frac{30}{100} = \frac{3}{10}$$

Percent Problems

The basic percent equation is the following:

$$\frac{is}{of} = \frac{\%}{100}$$

The placement of numbers in the equation depends on what the question asks.

Example 1

Find 40% of 80.

Basically, the problem is asking, "What is 40% of 80?" The 40% is the percent, and 80 is the number to find the percent "of." The equation is:

$$\frac{x}{80} = \frac{40}{100}$$

Solving the equation by cross-multiplication, the problem becomes $100x = 80(40)$. Solving for *x* produces the answer: $x = 32$.

Example 2

What percent of 100 is 20?

20 fills in the "is" portion, while 100 fills in the "of." The question asks for the percent, so that will be *x*, the unknown. The following equation is set up:

$$\frac{20}{100} = \frac{x}{100}$$

Cross-multiplying yields the equation $100x = 20(100)$. Solving for *x* gives the answer: 20%.

Example 3

30% of what number is 30?

The following equation uses the clues and numbers in the problem:

$$\frac{30}{x} = \frac{30}{100}$$

Cross-multiplying results in the equation $30(100) = 30x$. Solving for *x* gives the answer:$x = 100$.

Problems Involving Estimation

Sometimes when multiplying numbers, the result can be estimated by **rounding**. For example, to estimate the value of 11.2×2.01, each number can be rounded to the nearest integer. This will yield a result of 22.

Rate, Percent, and Measurement Problems

A **ratio** compares the size of one group to the size of another. For example, there may be a room with 4 tables and 24 chairs. The ratio of tables to chairs is $4: 24$. Such ratios behave like fractions in that both sides of the ratio by the same number can be multiplied or divided. Thus, the ratio 4:24 is the same as the ratio 2:12 and 1:6.

One quantity is **proportional** to another quantity if the first quantity is always some multiple of the second. For instance, the distance travelled in five hours is always five times to the speed as travelled. The distance is proportional to speed in this case.

One quantity is **inversely proportional** to another quantity if the first quantity is equal to some number divided by the second quantity. The time it takes to travel one hundred miles will be given by 100 divided by the speed travelled. The time is inversely proportional to the speed.

When dealing with word problems, there is no fixed series of steps to follow, but there are some general guidelines to use. It is important that the quantity to be found is identified. Then, it can be determined how the given values can be used and manipulated to find the final answer.

Example 1

Jana wants to travel to visit Alice, who lives one hundred and fifty miles away. If she can drive at fifty miles per hour, how long will her trip take?

The quantity to find is the *time* of the trip. The time of a trip is given by the distance to travel divided by the speed to be traveled. The problem determines that the distance is one hundred and fifty miles, while the speed is fifty miles per hour. Thus, 150 divided by 50 is $150 \div 50 = 3$. Because *miles* and *miles per hour* are the units being divided, the miles cancel out. The result is 3 hours.

Example 2

Bernard wishes to paint a wall that measures twenty feet wide by eight feet high. It costs ten cents to paint one square foot. How much money will Bernard need for paint?

The final quantity to compute is the *cost* to paint the wall. This will be ten cents ($0.10) for each square foot of area needed to paint. The area to be painted is unknown, but the dimensions of the wall are given; thus, it can be calculated.

The dimensions of the wall are 20 feet wide and 8 feet high. Since the area of a rectangle is length multiplied by width, the area of the wall is $8 \times 20 = 160$ square feet. Multiplying 0.1×160 yields $16 as the cost of the paint.

The **average** or **mean** of a collection of numbers is given by adding those numbers together and then dividing by the total number of values. A **weighted average** or **weighted mean** is given by adding the numbers multiplied by their weights, then dividing by the sum of the weights:

$$\frac{w_1x_1 + w_2x_2 + w_3x_3 \ldots + w_nx_n}{w_1 + w_2 + w_3 + \cdots + w_n}$$

An **ordinary average** is a weighted average where all the weights are 1.

Simple Geometry Problems

There are many key facts related to geometry that are applicable. The sum of the measures of the angles of a triangle are 180°, and for a quadrilateral, the sum is 360°. Rectangles and squares each have four right angles. A **right angle** has a measure of 90°.

Perimeter

The **perimeter** is the distance around a figure or the sum of all sides of a polygon.

The **formula for the perimeter of a square** is four times the length of a side. For example, the following square has side lengths of 8 feet:

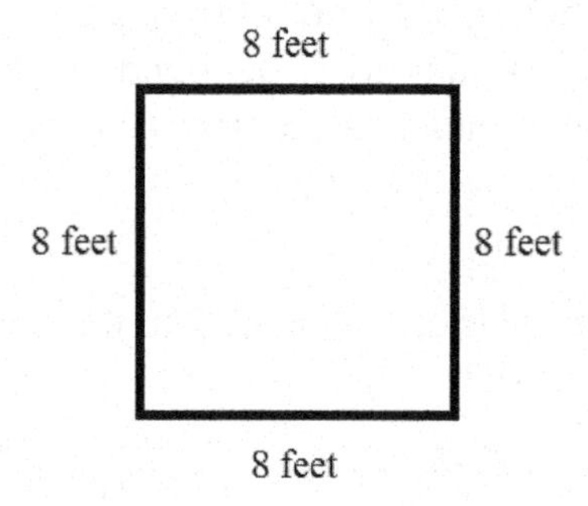

The perimeter is 32 feet because 4 times 8 is 32.

The **formula for a perimeter of a rectangle** is the sum of twice the length and twice the width. For example, if the length of a rectangle is 10 inches and the width 8 inches, then the perimeter is 36 inches because:

$$P = 2l + 2w$$

$$2(10) + 2(8)$$

$$20 + 16 = 36 \text{ inches}$$

Area

The area is the amount of space inside of a figure, and there are formulas associated with area.

The area of a triangle is the product of one-half the base and height. For example, if the base of the triangle is 2 feet and the height 4 feet, then the area is 4 square feet. The following equation shows the formula used to calculate the area of the triangle:

$$A = \frac{1}{2}bh = \frac{1}{2}(2)(4) = 4 \text{ square feet}$$

The area of a square is the length of a side squared. For example, if a side of a square is 7 centimeters, then the area is 49 square centimeters. The formula for this example is:

$$A = s^2 = 7^2 = 49 \text{ square centimeters}$$

An example is if the rectangle has a length of 6 inches and a width of 7 inches, then the area is 42 square inches:

$$A = lw = 6(7) = 42 \text{ square inches}$$

The area of a trapezoid is one-half the height times the sum of the bases. For example, if the length of the bases are 2.5 and 3 feet and the height 3.5 feet, then the area is 9.625 square feet. The following formula shows how the area is calculated:

$$A = \frac{1}{2}h(b_1 + b_2)$$

$$\frac{1}{2}(3.5)(2.5 + 3)$$

$$\frac{1}{2}(3.5)(5.5) = 9.625 \text{ square feet}$$

The perimeter of a figure is measured in single units, while the area is measured in square units.

Distribution of a Quantity into its Fractional Parts

A quantity may be broken into its fractional parts. For example, a toy box holds three types of toys for kids. $\frac{1}{3}$ of the toys are Type A and $\frac{1}{4}$ of the toys are Type B. With that information, how many Type C toys are there?

First, the sum of Type A and Type B must be determined by finding a common denominator to add the fractions. The lowest common multiple is 12, so that is what will be used. The sum is:

$$\frac{1}{3} + \frac{1}{4} = \frac{4}{12} + \frac{3}{12} = \frac{7}{12}$$

This value is subtracted from 1 to find the number of Type C toys. The value is subtracted from 1 because 1 represents a whole. The calculation is:

$$1 - \frac{7}{12} = \frac{12}{12} - \frac{7}{12} = \frac{5}{12}$$

This means that $\frac{5}{12}$ of the toys are Type C. To check the answer, add all fractions together, and the result should be 1.

Algebra

Solving for X in Proportions

Proportions are commonly used to solve word problems to find unknown values such as *x* that are some percent or fraction of a known number. Proportions are solved by cross-multiplying and then dividing to arrive at x. The following examples show how this is done:

1) $\frac{75\%}{90\%} = \frac{25\%}{x}$

To solve for x, the fractions must be cross multiplied:

$$75\%x = 90\% \times 25\%$$

To make things easier, let's convert the percentages to decimals:

$$0.9 \times 0.25 = 0.225 = 0.75x$$

To get rid of x's coefficient, each side must be divided by that same coefficient to get the answer $x = 0.3$. The question could ask for the answer as a percentage or fraction in lowest terms, which are 30% and $\frac{3}{10}$, respectively.

2) $\frac{x}{12} = \frac{30}{96}$

Cross-multiply: $96x = 30 \times 12$
Multiply: $96x = 360$
Divide: $x = 360 \div 96$
Answer: $x = 3.75$

3) $\frac{0.5}{3} = \frac{x}{6}$

Cross-multiply: $3x = 0.5 \times 6$
Multiply: $3x = 3$
Divide: $x = 3 \div 3$
Answer: $x = 1$

You may have noticed there's a faster way to arrive at the answer. If there is an obvious operation being performed on the proportion, the same operation can be used on the other side of the proportion to solve for x. For example, in the first practice problem, 75% became 25% when divided by 3, and upon doing the same to 90%, the correct answer of 30% would have been found with much less legwork. However, these questions aren't always so intuitive, so it's a good idea to work through the steps, even if the answer seems apparent from the outset.

Translating Words into Math

To translate a word problem into an expression, look for a series of key words indicating addition, subtraction, multiplication, or division:

Addition: *add, altogether, together, plus, increased by, more than, in all, sum,* and *total*

Subtraction: *minus, less than, difference, decreased by, fewer than, remain,* and *take away*

Multiplication: *times, twice, of, double*, and *triple*

Division: *divided by, cut up, half, quotient of, split,* and *shared equally*

If a question asks to give words to a mathematical expression and says "equals," then an = sign must be included in the answer. Similarly, "less than or equal to" is expressed by the inequality symbol ≤, and "greater than or equal" to is expressed as ≥. Furthermore, "less than" is represented by <, and "greater than" is expressed by >.

Word Problems

Word problems can appear daunting, but don't let the verbiage psych you out. No matter the scenario or specifics, the key to answering them is to translate the words into a math problem. Always keep in mind what the question is asking and what operations could lead to that answer. The following word problem resembles one of the question types most frequently encountered on the exam.

Working with Money

Walter's Coffee Shop sells a variety of drinks and breakfast treats.

Price List	
Hot Coffee	$2.00
Slow-Drip Iced Coffee	$3.00
Latte	$4.00
Muffin	$2.00
Crepe	$4.00
Egg Sandwich	$5.00

Costs	
Hot Coffee	$0.25
Slow-Drip Iced Coffee	$0.75
Latte	$1.00
Muffin	$1.00
Crepe	$2.00
Egg Sandwich	$3.00

Walter's utilities, rent, and labor costs him $500 per day. Today, Walter sold 200 hot coffees, 100 slow-drip iced coffees, 50 lattes, 75 muffins, 45 crepes, and 60 egg sandwiches. What was Walter's total profit today?

To accurately answer this type of question, determine the total cost of making his drinks and treats, then determine how much revenue he earned from selling those products. After arriving at these two totals, the profit is measured by deducting the total cost from the total revenue.

Walter's costs for today:

Item	Quantity	Cost Per Unit	Total Cost
Hot Coffee	200	$0.25	$50
Slow-Drip Iced Coffee	100	$0.75	$75
Latte	50	$1.00	$50
Muffin	75	$1.00	$75
Crepe	45	$2.00	$90
Egg Sandwich	60	$3.00	$180
Utilities, rent, and labor			$500
Total Costs			$1,020

Walter's revenue for today:

Item	Quantity	Revenue Per Unit	Total Revenue
Hot Coffee	200	$2.00	$400
Slow-Drip Iced Coffee	100	$3.00	$300
Latte	50	$4.00	$200
Muffin	75	$2.00	$150
Crepe	45	$4.00	$180
Egg Sandwich	60	$5.00	$300
Total Revenue			$1,530

$$\text{Walter's Profit} = \text{Revenue} - \text{Costs} = \$1{,}530 - \$1{,}020 = \$510$$

This strategy is applicable to other question types. For example, calculating salary after deductions, balancing a checkbook, and calculating a dinner bill are common word problems similar to business planning. Just remember to use the correct operations. When a balance is increased, use addition. When a balance is decreased, use subtraction. Common sense and organization are your greatest assets when answering word problems.

Unit Rate

Unit rate word problems will ask to calculate the rate or quantity of something in a different value. For example, a problem might say that a car drove a certain number of miles in a certain number of minutes and then ask how many miles per hour the car was traveling. These questions involve solving proportions. Consider the following examples:

1) Alexandra made $96 during the first 3 hours of her shift as a temporary worker at a law office. She will continue to earn money at this rate until she finishes in 5 more hours. How much does Alexandra make per hour? How much will Alexandra have made at the end of the day?

This problem can be solved in two ways. The first is to set up a proportion, as the rate of pay is constant. The second is to determine her hourly rate, multiply the 5 hours by that rate, and then add the $96.

To set up a proportion, put the money already earned over the hours already worked on one side of an equation. The other side has x over 8 hours (the total hours worked in the day). It looks like this:

$$\frac{96}{3} = \frac{x}{8}$$

Now, cross-multiply to get $768 = 3x$. To get x, divide by 3, which leaves $x = 256$. Alternatively, as x is the numerator of one of the proportions, multiplying by its denominator will reduce the solution by one step. Thus, Alexandra will make $256 at the end of the day. To calculate her hourly rate, divide the total by 8, giving $32 per hour.

Alternatively, it is possible to figure out the hourly rate by dividing $96 by 3 hours to get $32 per hour. Now her total pay can be figured by multiplying $32 per hour by 8 hours, which comes out to $256.

2) Jonathan is reading a novel. So far, he has read 215 of the 335 total pages. It takes Jonathan 25 minutes to read 10 pages, and the rate is constant. How long does it take Jonathan to read one page? How much longer will it take him to finish the novel? Express the answer in time.

To calculate how long it takes Jonathan to read one page, divide the 25 minutes by 10 pages to determine the page per minute rate. Thus, it takes 2.5 minutes to read one page.

Jonathan must read 120 more pages to complete the novel. (This is calculated by subtracting the pages already read from the total.) Now, multiply his rate per page by the number of pages. Thus,

$$120 \times 2.5 = 300$$

Expressed in time, 300 minutes is equal to 5 hours.

3) At a hotel, $\frac{4}{5}$ of the 120 rooms are booked for Saturday. On Sunday, $\frac{3}{4}$ of the rooms are booked. On which day are more of the rooms booked, and by how many more?

The first step is to calculate the number of rooms booked for each day. Do this by multiplying the fraction of the rooms booked by the total number of rooms.

Saturday: $\frac{4}{5} \times 120 = \frac{4}{5} \times \frac{120}{1} = \frac{480}{5} = 96 \text{ rooms}$

Sunday: $\frac{3}{4} \times 120 = \frac{3}{4} \times \frac{120}{1} = \frac{360}{4} = 90 \text{ rooms}$

Thus, more rooms were booked on Saturday by 6 rooms.

4) In a veterinary hospital, the veterinarian-to-pet ratio is 1:9. The ratio is always constant. If there are 45 pets in the hospital, how many veterinarians are currently in the veterinary hospital?

Set up a proportion to solve for the number of veterinarians:

$$\frac{1}{9} = \frac{x}{45}$$

Cross-multiplying results in $9x = 45$, which works out to 5 veterinarians.

Alternatively, as there are always 9 times as many pets as veterinarians, it is possible to divide the number of pets (45) by 9. This also arrives at the correct answer of 5 veterinarians.

5) At a general practice law firm, 30% of the lawyers work solely on tort cases. If 9 lawyers work solely on tort cases, how many lawyers work at the firm?

First, solve for the total number of lawyers working at the firm, which will be represented here with x. The problem states that 9 lawyers work solely on torts cases, and they make up 30% of the total lawyers at the firm. Thus, 30% multiplied by the total, x, will equal 9. Written as equation, this is:

$$30\% \times x = 9$$

It's easier to deal with the equation after converting the percentage to a decimal, leaving $0.3x = 9$. Thus, $x = \frac{9}{0.3} = 30$ lawyers working at the firm.

6) Xavier was hospitalized with pneumonia. He was originally given 35mg of antibiotics. Later, after his condition continued to worsen, Xavier's dosage was increased to 60mg. What was the percent increase of the antibiotics? Round the percentage to the nearest tenth.

An increase or decrease in percentage can be calculated by dividing the difference in amounts by the original amount and multiplying by 100. Written as an equation, the formula is:

$$\frac{\text{new quantity} - \text{old quantity}}{\text{old quantity}} \times 100$$

Here, the question states that the dosage was increased from 35mg to 60mg, so these are plugged into the formula to find the percentage increase.

$$\frac{60-35}{35} \times 100 = \frac{25}{35} \times 100$$

$$0.7142 \times 100 = 71.4\%$$

Order of Rational Numbers

A common question type asks to order rational numbers from least to greatest or greatest to least. The numbers will come in a variety of formats, including decimals, percentages, roots, fractions, and whole numbers. These questions test for knowledge of different types of numbers and the ability to determine their respective values.

Whether the question asks to order the numbers from greatest to least or least to greatest, the crux of the question is the same—convert the numbers into a common format. Generally, it's easiest to write the numbers as whole numbers and decimals so they can be placed on a number line. Follow these examples to understand this strategy.

1) Order the following rational numbers from greatest to least:

$$\sqrt{36}, 0.65, 78\%, \frac{3}{4}, 7, 90\%, \frac{5}{2}$$

Of the seven numbers, the whole number (7) and decimal (0.65) are already in an accessible form, so concentrate on the other five.

First, the square root of 36 equals 6. (If the test asks for the root of a non-perfect root, determine which two whole numbers the root lies between.) Next, convert the percentages to decimals. A percentage means "per hundred," so this conversion requires moving the decimal point two places to the left, leaving 0.78 and 0.9. Lastly, evaluate the fractions:

$$\frac{3}{4} = \frac{75}{100} = 0.75\,; \frac{5}{2} = 2\frac{1}{2} = 2.5$$

Now, the only step left is to list the numbers in the request order:

$$7, \sqrt{36}, \frac{5}{2}, 90\%, 78\%, \frac{3}{4}, 0.65$$

2) Order the following rational numbers from least to greatest:

$$2.5, \sqrt{9}, -10.5, 0.853, 175\%, \sqrt{4}, \frac{4}{5}$$

$$\sqrt{9} = 3$$

$$175\% = 1.75$$

$$\sqrt{4} = 2$$

$$\frac{4}{5} = 0.8$$

From least to greatest, the answer is: $-10.5, \frac{4}{5}, 0.853, 175\%, \sqrt{4}, 2.5, \sqrt{9},$

FOIL Method

FOIL is a technique for generating polynomials through the multiplication of binomials. A polynomial is an expression of multiple variables (for example, x, y, z) in at least three terms involving only the four basic operations and exponents. FOIL is an acronym for First, Outer, Inner, and Last. "First" represents the multiplication of the terms appearing first in the binomials. "Outer" means multiplying the outermost terms. "Inner" means multiplying the terms inside. "Last" means multiplying the last terms of each binomial.

After completing FOIL and solving the operations, like terms are combined. To identify like terms, look for terms with the same variable and the same exponent. For example, look at:

$$4x^2 - x^2 + 15x + 2x^2 - 8$$

The $4x^2$, $-x^2$, and $2x^2$ are all like terms because they have the variable (x) and exponent (2). Thus, after combining the like terms, the polynomial has been simplified:

$$5x^2 + 15x - 8$$

The purpose of FOIL is to simplify an equation involving multiple variables and operations. Although it sounds complicated, working through some examples will provide some clarity:

1) Simplify $(x + 10)(x + 4) = (x \times x) + (x \times 4) + (10 \times x) + (10 \times 4)$

First Outer Inner Last

After multiplying these binomials, it's time to solve the operations and combine like terms. Thus, the expression becomes:

$$x^2 + 4x + 10x + 40 = x^2 + 14x + 40$$

2) Simplify $2x\ (4x^3 - 7y^2 + 3x^2 + 4)$

Here, a monomial ($2x$) is multiplied into a polynomial:

$$(4x^3 - 7y^2 + 3x^2 + 4)$$

Using the distributive property, multiply the monomial against each term in the polynomial. This becomes:

$$2x(4x^3) - 2x(7y^2 + 2x(3x^2) + 2x(4)\ .$$

Now, simplify each monomial. Start with the coefficients:

$$(2 \times 4)(x \times x^3) - (2 \times 7)(x \times y^2) + (2 \times 3)(x \times x^2) + (2 \times 4)(x)$$

When multiplying powers with the same base, add their exponents. Remember, a variable with no listed exponent has an exponent of 1, and exponents of distinct variables cannot be combined. This produces the answer:

$$8x^{1+3} - 14xy^2 + 6x^{1+2} + 8x = 8x^4 - 14xy^2 + 6x^3 + 8x$$

3) Simplify $(8x^{10}y^2z^4) \div (4x^2y^4z^7)$

First, divide the coefficients of the first two polynomials: $8 \div 4 = 2$. Second, divide exponents with the same variable, which requires subtracting the exponents. This results in:

$$2(x^{10-2}y^{2-4}z^{4-7}) = 2x^8y^{-2}z^{-3}$$

However, the most simplified answer should include only positive exponents. Thus, $y^{-2}z^{-3}$ needs to be converted into fractions, respectively $\frac{1}{y^2}$ and $\frac{1}{z^3}$. Since the $2x^8$ has a positive exponent, it is placed in the numerator, and $\frac{1}{y^2}$ and $\frac{1}{z^3}$ are combined into the denominator, leaving $\frac{2x^8}{y^2z^3}$ as the final answer.

Geometry

Plane Geometry

Locations on the plane that have no width or breadth are called **points**. These points usually will be denoted with capital letters such as *P*.

Any pair of points *A*, *B* on the plane will determine a unique straight line between them. This line is denoted *AB*. Sometimes to emphasize a line is being considered, this will be written as $\overleftrightarrow{AB}$.

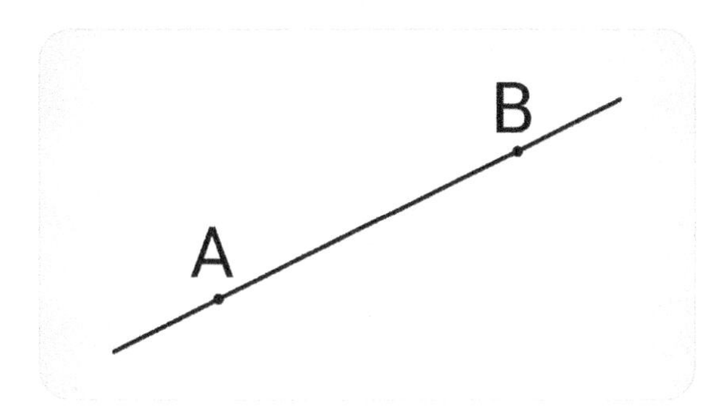

If the Cartesian coordinates for *A* and *B* are known, then the distance $d(A, B)$ along the line between them can be measured using the **Pythagorean formula**, which states that if $A = (x_1, y_1)$ and $B = (x_2, y_2)$, then the distance between them is:

$$d(A, B) = \sqrt{(x_2 - x_1)^2 + (y_2 - y_1)^2}$$

The part of a line that lies between *A* and *B* is called a **line segment**. It has two endpoints, one at *A* and one at *B*. **Rays** also can be formed. Given points *A* and *B*, a **ray** is the portion of a line that starts at one of these points, passes through the other, and keeps on going. Therefore, a ray has a single endpoint, but the other end goes off to infinity.

Given a pair of points *A* and *B*, a circle centered at *A* and passing through *B* can be formed. This is the set of points whose distance from *A* is exactly $d(A, B)$. The radius of this circle will be $d(A, B)$.

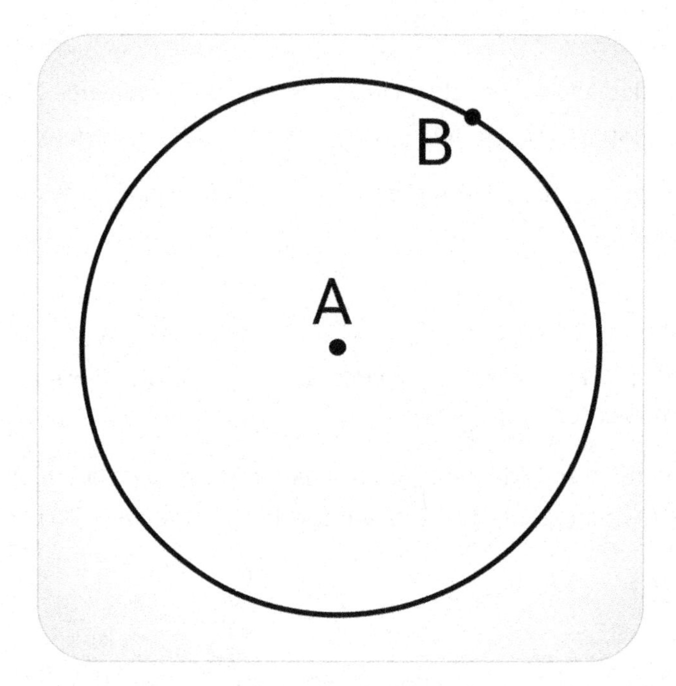

The **circumference** of a circle is the distance traveled by following the edge of the circle for one complete revolution, and the length of the circumference is given by $2\pi r$, where *r* is the radius of the circle. The formula for circumference is $C = 2\pi r$.

When two lines cross, they form an **angle**. The point where the lines cross is called the **vertex** of the angle.

The angle can be named by either just using the vertex, $\angle A$, or else by listing three points $\angle BAC$, as shown in the diagram below.

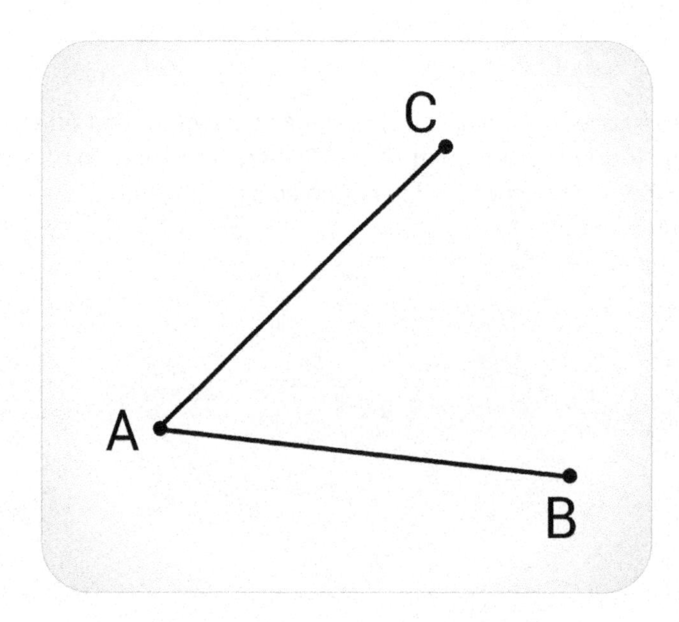

The measurement of an angle can be given in degrees or in radians. In degrees, a full circle is 360 degrees, written 360°. In radians, a full circle is 2π radians.

Given two points on the circumference of a circle, the path along the circle between those points is called an **arc** of the circle. For example, the arc between *B* and *C* is denoted by a thinner line:

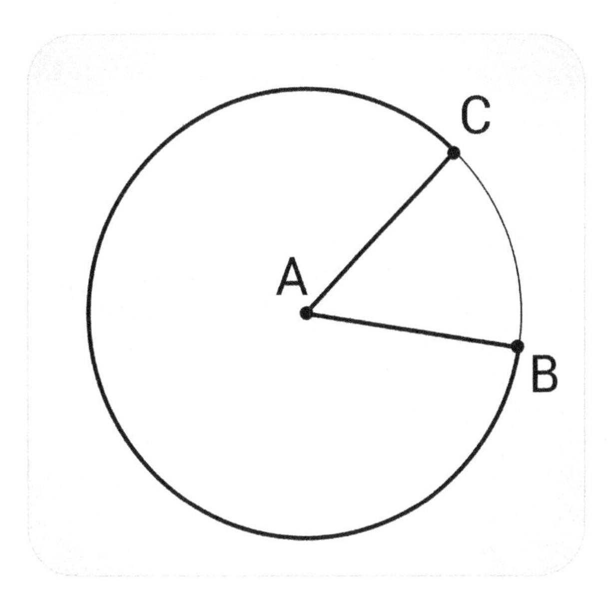

The length of the path along an arc is called the **arc length**. If the circle has radius *r*, then the arc length is given by multiplying the measure of the angle in radians by the radius of the circle.

Two lines are said to be **parallel** if they never intersect. If the lines are *AB* and *CD*, then this is written as $AB \parallel CD$.

If two lines cross to form four quarter-circles, that is, 90° angles, the two lines are **perpendicular**. If the point at which they cross is *B*, and the two lines are *AB* and *BC*, then this is written as $AB \perp BC$.

A **polygon** is a closed figure (meaning it divides the plane into an inside and an outside) consisting of a collection of line segments between points. These points are called the **vertices** of the polygon. These line segments must not overlap one another. Note that the number of sides is equal to the number of angles, or vertices of the polygon. The angles between line segments meeting one another in the polygon are called **interior angles**.

A **regular polygon** is a polygon whose edges are all the same length and whose interior angles are all of equal measure.

A **triangle** is a polygon with three sides. A **quadrilateral** is a polygon with four sides.

A **right triangle** is a triangle that has one 90° angle.

The sum of the interior angles of any triangle must add up to 180°.

An **isosceles triangle** is a triangle in which two of the sides are the same length. In this case, it will always have two congruent interior angles. If a triangle has two congruent interior angles, it will always be isosceles.

An **equilateral triangle** is a triangle whose sides are all the same length and whose angles are all equivalent to one another, equal to 60°. Equilateral triangles are examples of regular polygons. Note that equilateral triangles are also isosceles.

A **rectangle** is a quadrilateral whose interior angles are all 90°. A rectangle has two sets of sides that are equal to one another.

A **square** is a rectangle whose width and height are equal. Therefore, squares are regular polygons.

A **parallelogram** is a quadrilateral in which the opposite sides are parallel and equivalent to each other.

Transformations of a Plane

Given a figure drawn on a plane, many changes can be made to that figure, including **rotation**, **translation**, and **reflection**. Rotations turn the figure about a point, translations slide the figure, and reflections flip the figure over a specified line. When performing these transformations, the original figure is called the **pre-image**, and the figure after transformation is called the **image**.

More specifically, **translation** means that all points in the figure are moved in the same direction by the same distance. In other words, the figure is slid in some fixed direction. Of course, while the entire figure is slid by the same distance, this does not change any of the measurements of the figures involved. The result will have the same distances and angles as the original figure.

In terms of Cartesian coordinates, a translation means a shift of each of the original points (x, y) by a fixed amount in the *x* and *y* directions, to become $(x + a, y + b)$.

Another procedure that can be performed is called **reflection**. To do this, a line in the plane is specified, called the **line of reflection**. Then, take each point and flip it over the line so that it is the same distance from the line but on the opposite side of it. This does not change any of the distances or angles involved, but it does reverse the order in which everything appears.

To reflect something over the *x*-axis, the points (x, y) are sent to $(x, -y)$. To reflect something over the *y*-axis, the points (x, y) are sent to the points $(-x, y)$. Flipping over other lines is not something easy to express in Cartesian coordinates. However, by drawing the figure and the line of reflection, the distance to the line and the original points can be used to find the reflected figure.

Example: Reflect this triangle with vertices (-1, 0), (2, 1), and (2, 0) over the *y*-axis. The pre-image is shown below.

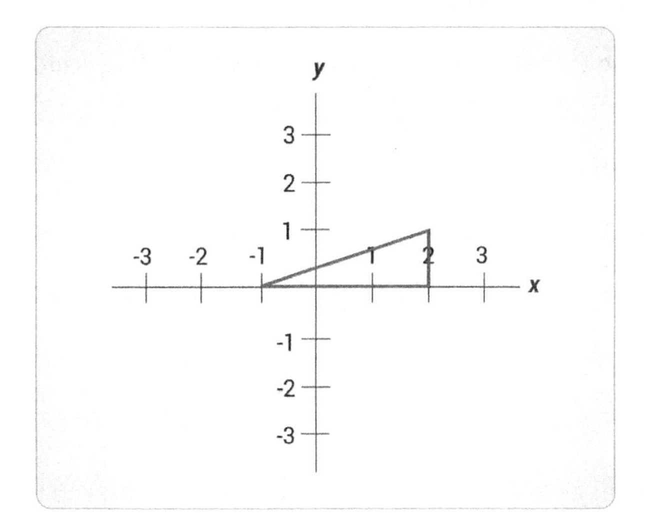

To do this, flip the *x* values of the points involved to the negatives of themselves, while keeping the *y* values the same. The image is shown here.

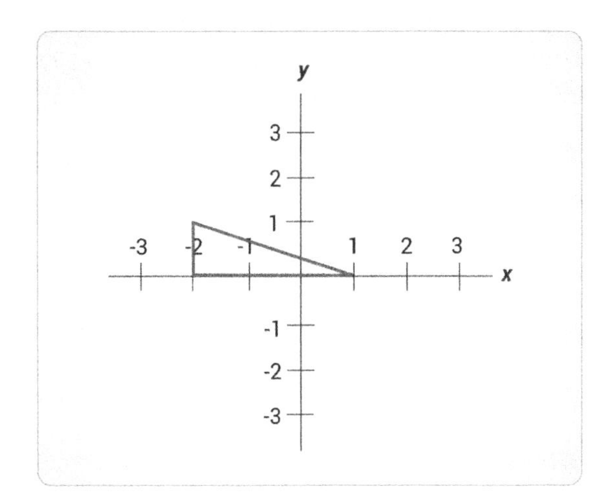

The new vertices will be (1, 0), (-2, 1), and (-2, 0).

Another procedure that does not change the distances and angles in a figure is **rotation**. In this procedure, pick a center point, then rotate every vertex along a circle around that point by the same angle. This procedure is also not easy to express in Cartesian coordinates, and this is not a requirement on this test. However, as with reflections, it's helpful to draw the figures and see what the result of the rotation would look like. This transformation can be performed using a compass and protractor.

Each one of these transformations can be performed on the coordinate plane without changes to the original dimensions or angles.

If two figures in the plane involve the same distances and angles, they are called **congruent figures**. In other words, two figures are congruent when they go from one form to another through reflection, rotation, and translation, or a combination of these.

Remember that rotation and translation will give back a new figure that is identical to the original figure, but reflection will give back a mirror image of it.

To recognize that a figure has undergone a rotation, check to see that the figure has not been changed into a mirror image, but that its orientation has changed (that is, whether the parts of the figure now form different angles with the *x* and *y* axes).

To recognize that a figure has undergone a translation, check to see that the figure has not been changed into a mirror image, and that the orientation remains the same.

To recognize that a figure has undergone a reflection, check to see that the new figure is a mirror image of the old figure.

Keep in mind that sometimes a combination of translations, reflections, and rotations may be performed on a figure.

Dilation

A **dilation** is a transformation that preserves angles, but not distances. This can be thought of as stretching or shrinking a figure. If a dilation makes figures larger, it is called an **enlargement**. If a dilation makes figures smaller, it is called a **reduction**. The easiest example is to dilate around the origin. In this case, multiply the *x* and *y* coordinates by a **scale factor**, k, sending points (x, y) to (kx, ky).

As an example, draw a dilation of the following triangle, whose vertices will be the points (-1, 0), (1, 0), and (1, 1).

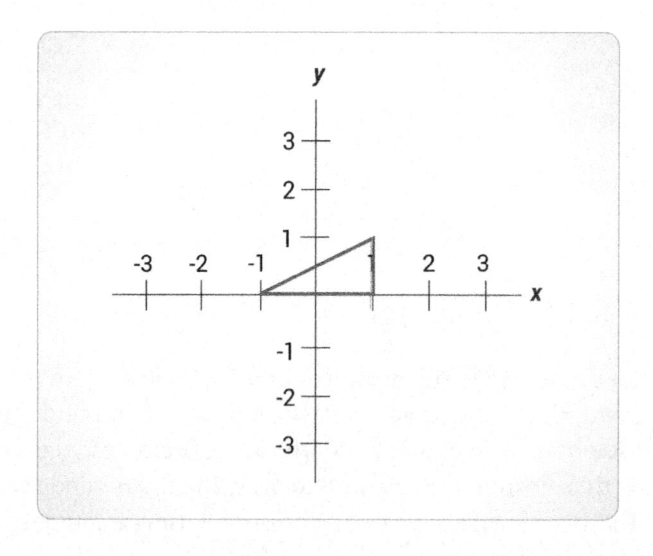

For this problem, dilate by a scale factor of 2, so the new vertices will be (-2, 0), (2, 0), and (2, 2).

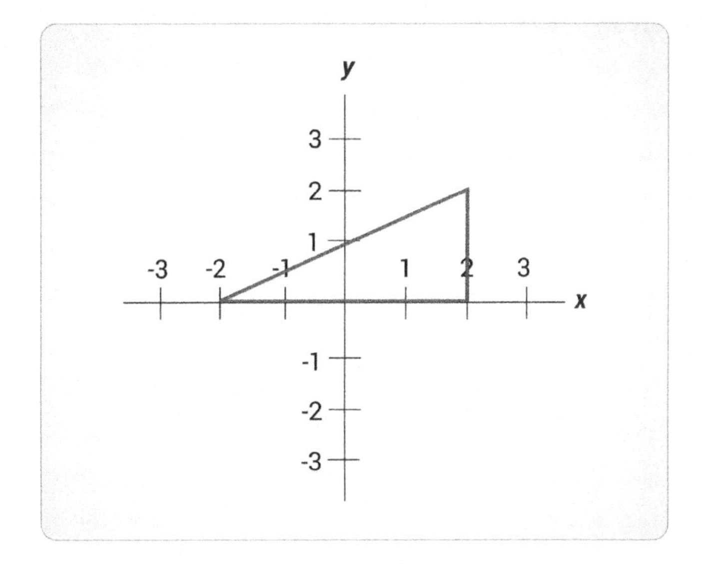

Note that after a dilation, the distances between the vertices of the figure will have changed, but the angles will remain the same. The two figures that are obtained by dilation, along with possibly translation, rotation, and reflection, are all **similar** to one another. Another way to think of this is that similar figures have the same number of vertices and edges, and their angles are all the same. Similar figures have the same basic shape but are different in size.

Symmetry

Using the types of transformations above, if an object can undergo these changes and not appear to have changed, then the figure is symmetrical. If an object can be split in half by a line and flipped over that line to lie directly on top of itself, it is said to have **line symmetry**. An example of both types of figures is seen below.

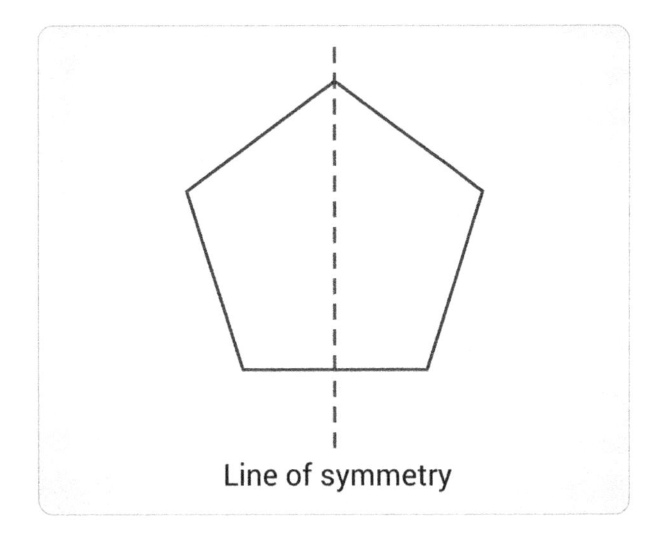

If an object can be rotated about its center to any degree smaller than 360, and it lies directly on top of itself, the object is said to have **rotational symmetry**. An example of this type of symmetry is shown below. The pentagon has an order of 5.

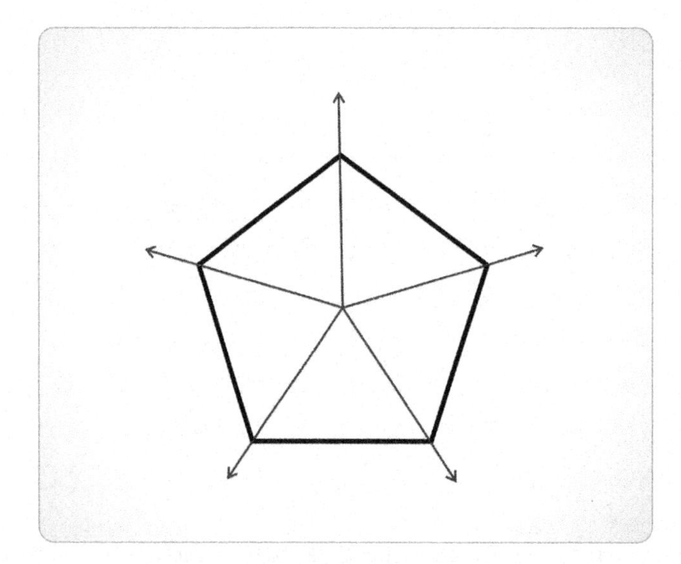

The rotational symmetry lines in the figure above can be used to find the angles formed at the center of the pentagon. Knowing that all of the angles together form a full circle, at 360 degrees, the figure can be split into 5 angles equally. By dividing the 360° by 5, each angle is 72°.

Given the length of one side of the figure, the perimeter of the pentagon can also be found using rotational symmetry. If one side length was 3 cm, that side length can be rotated onto each other side length four times. This would give a total of 5 side lengths equal to 3 cm. To find the perimeter, or distance around the figure, multiply 3 by 5. The perimeter of the figure would be 15 cm.

If a line cannot be drawn anywhere on the object to flip the figure onto itself or rotated less than or equal to 180 degrees to lay on top of itself, the object is asymmetrical. Examples of these types of figures are shown below.

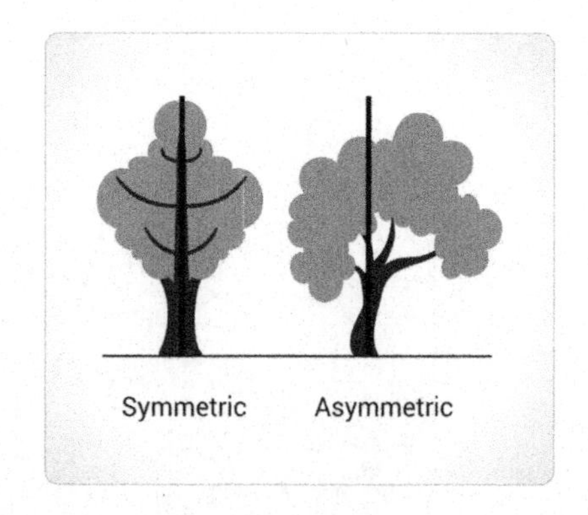

Perimeters and Areas

The **perimeter** of a polygon is the total length of a trip around the whole polygon, starting and ending at the same point. It is found by adding up the lengths of each line segment in the polygon. For a rectangle with sides of length *x* and *y*, the perimeter will be $2x + 2y$.

The area of a polygon is the area of the region that it encloses. Regarding the area of a rectangle with sides of length *x* and *y*, the area is given by xy. For a triangle with a base of length *b* and a height of length *h*, the area is: $\frac{1}{2}bh$

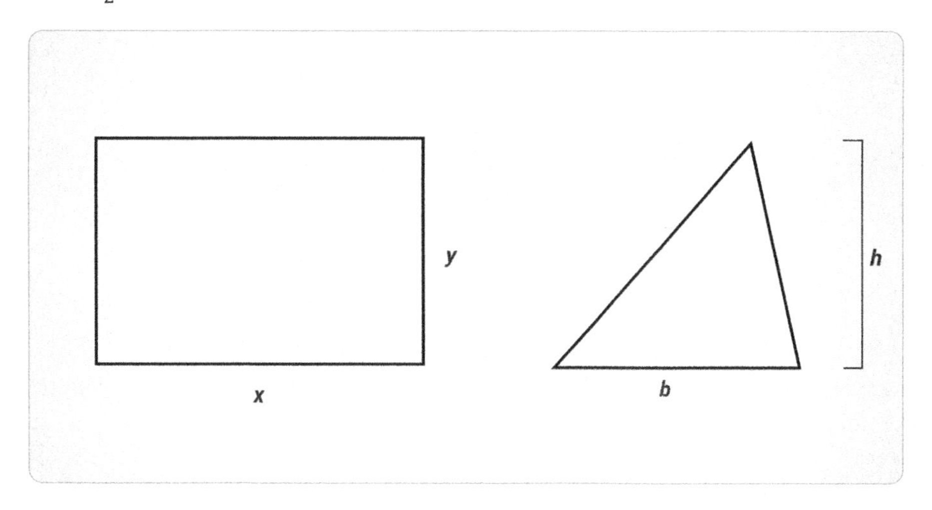

To find the areas of more general polygons, it is usually easiest to break up the polygon into rectangles and triangles. For example, find the area of the following figure whose vertices are (-1, 0), (-1, 2), (1, 3), and (1, 0).

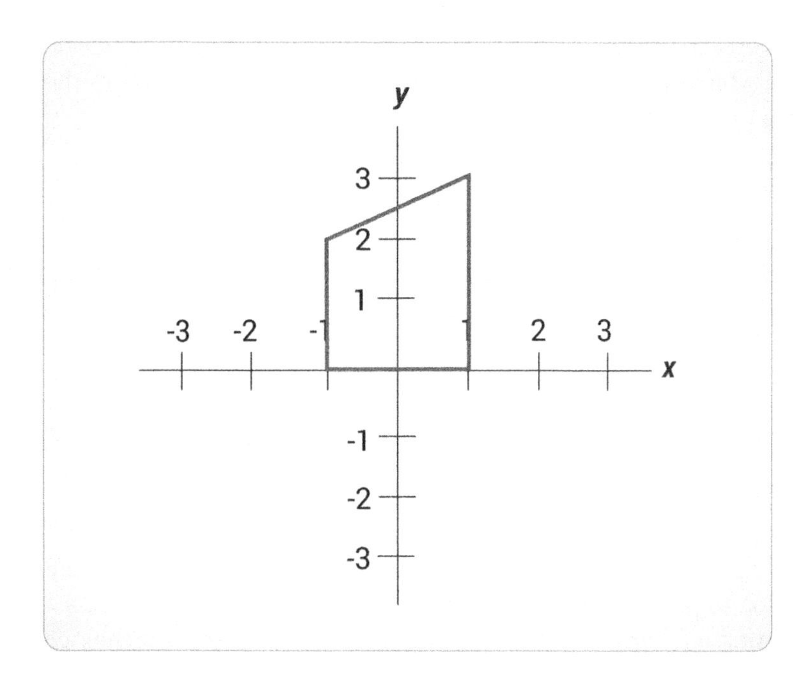

Separate this into a rectangle and a triangle as shown:

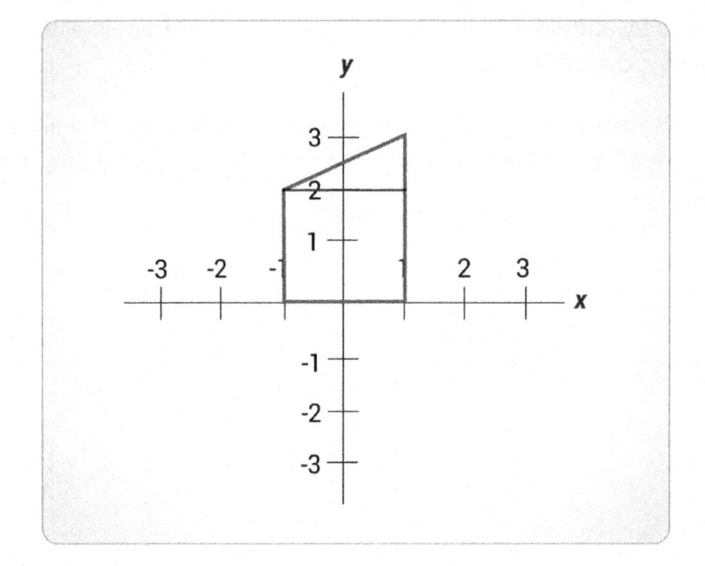

The rectangle has a height of 2 and a width of 2, so it has a total area of $2 \times 2 = 4$. The triangle has a width of 2 and a height of 1, so it has an area of:

$$\frac{1}{2} 2 \times 1 = 1$$

Therefore, the entire quadrilateral has an area of $4 + 1 = 5$.

As another example, suppose someone wants to tile a rectangular room that is 10 feet by 6 feet using triangular tiles that are 12 inches by 6 inches. How many tiles would be needed? To figure this, first find the area of the room, which will be $10 \times 6 = 60$ square feet. The dimensions of the triangle are 1 foot by ½ foot, so the area of each triangle is:

$$\frac{1}{2} \times 1 \times \frac{1}{2} = \frac{1}{4} \text{ square feet}$$

Notice that the dimensions of the triangle had to be converted to the same units as the rectangle. Now, take the total area divided by the area of one tile to find the answer:

$$\frac{60}{\frac{1}{4}} = 60 \times 4 = 240 \text{ tiles required}$$

Volumes and Surface Areas

Geometry in three dimensions is similar to geometry in two dimensions. The main new feature is that three points now define a unique **plane** that passes through each of them. Three dimensional objects can be made by putting together two-dimensional figures in different surfaces. Below, some of the possible three-dimensional figures will be provided, along with formulas for their volumes and surface areas.

A rectangular prism is a box whose sides are all rectangles meeting at 90° angles. Such a box has three dimensions: length, width, and height. If the length is *x*, the width is *y*, and the height is *z*, then the volume is given by $V = xyz$.

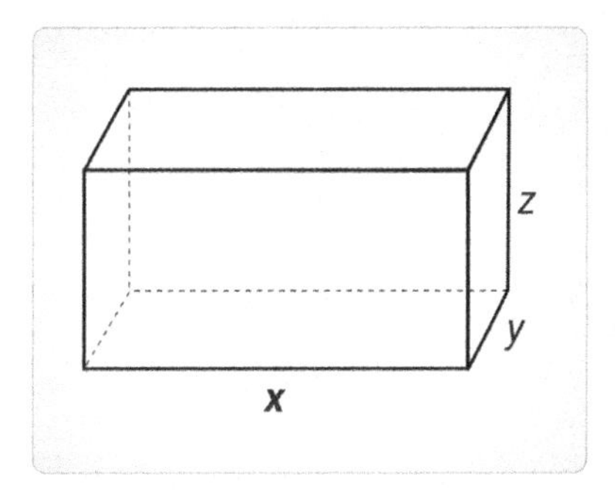

The surface area will be given by computing the surface area of each rectangle and adding them together. There are a total of six rectangles. Two of them have sides of length *x* and *y*, two have sides of length *y* and *z*, and two have sides of length *x* and *z*. Therefore, the total surface area will be given by:

$$SA = 2xy + 2yz + 2xz$$

A **rectangular pyramid** is a figure with a rectangular base and four triangular sides that meet at a single vertex. If the rectangle has sides of length *x* and *y*, then the volume will be given by:

$$V = \frac{1}{3}xyh$$

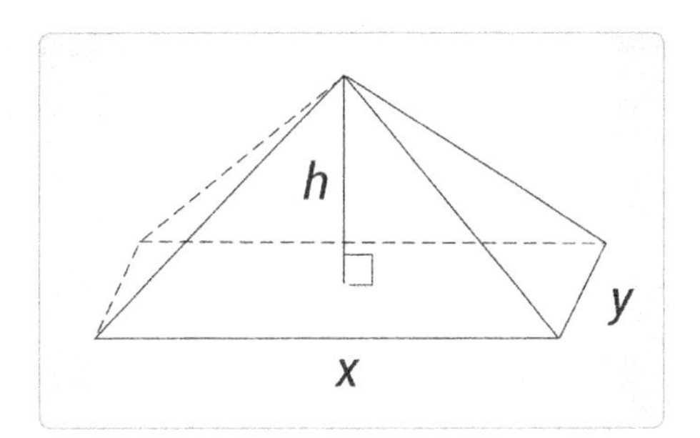

To find the surface area, the dimensions of each triangle need to be known. However, these dimensions can differ depending on the problem in question. Therefore, there is no general formula for calculating total surface area.

A **sphere** is a set of points all of which are equidistant from some central point. It is like a circle, but in three dimensions. The volume of a sphere of radius *r* is given by $V = \frac{4}{3}\pi r^3$. The surface area is given by $A = 4\pi r^2$.

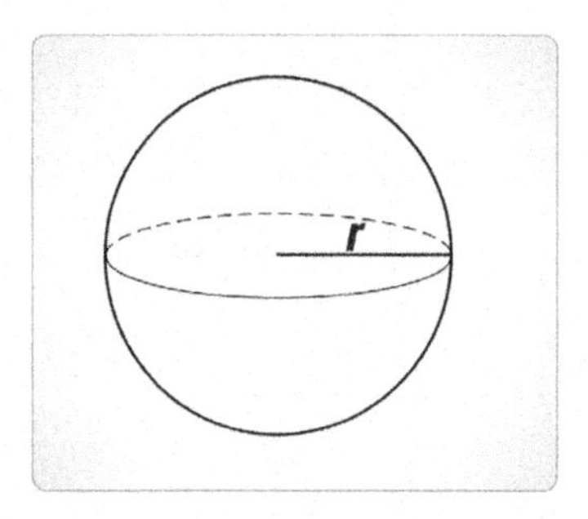

The Pythagorean Theorem

The Pythagorean theorem is an important concept in geometry. It states that for right triangles, the sum of the squares of the two shorter sides will be equal to the square of the longest side (also called the **hypotenuse**). The longest side will always be the side opposite to the 90° angle. If this side is called *c*, and the other two sides are *a* and *b*, then the Pythagorean theorem states that:

$$c^2 = a^2 + b^2$$

Since lengths are always positive, this also can be written as:

$$c = \sqrt{a^2 + b^2}$$

A diagram to show the parts of a triangle using the Pythagorean theorem is below.

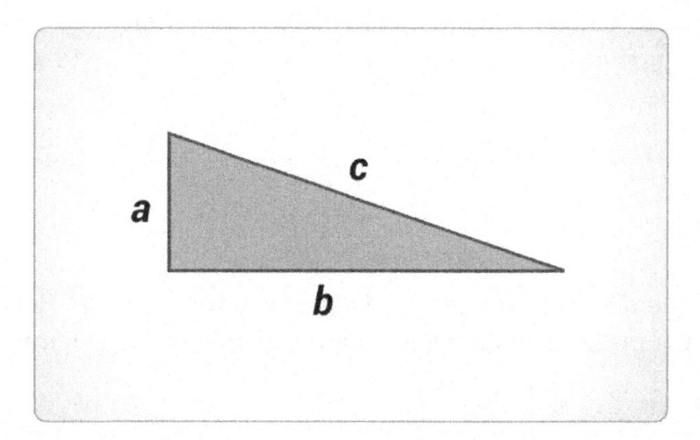

As an example of the theorem, suppose that Shirley has a rectangular field that is 5 feet wide and 12 feet long, and she wants to split it in half using a fence that goes from one corner to the opposite corner. How long will this fence need to be? To figure this out, note that this makes the field into two right triangles, whose hypotenuse will be the fence dividing it in half. Therefore, the fence length will be given by:

$$\sqrt{5^2 + 12^2} = \sqrt{169} = 13 \text{ feet long}$$

Similar Figures and Proportions

Sometimes, two figures are similar, meaning they have the same basic shape and the same interior angles, but they have different dimensions. If the ratio of two corresponding sides is known, then that ratio, or scale factor, holds true for all of the dimensions of the new figure.

Here is an example of applying this principle. Suppose that Lara is 5 feet tall and is standing 30 feet from the base of a light pole, and her shadow is 6 feet long. How high is the light on the pole? To figure this, it helps to make a sketch of the situation:

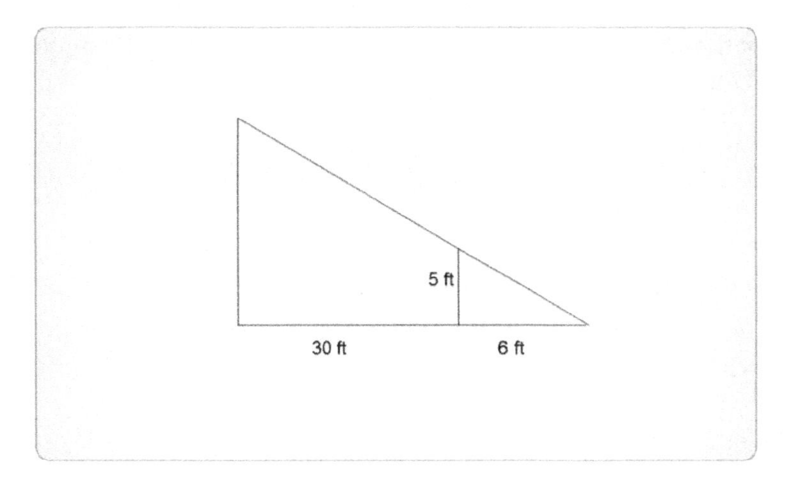

The light pole is the left side of the triangle. Lara is the 5-foot vertical line. Notice that there are two right triangles here, and that they have all the same angles as one another. Therefore, they form similar triangles. So, the ratio of proportionality between them must be determined.

The bases of these triangles are known. The small triangle, formed by Lara and her shadow, has a base of 6 feet. The large triangle formed by the light pole along with the line from the base of the pole out to the end of Lara's shadow is $30 + 6 = 36$ feet long. So, the ratio of the big triangle to the little triangle will be $\frac{36}{6} = 6$. The height of the little triangle is 5 feet. Therefore, the height of the big triangle will be $6 \times 5 = 30$ feet, meaning that the light is 30 feet up the pole.

Notice that the perimeter of a figure changes by the ratio of proportionality between two similar figures, but the area changes by the **square** of the ratio. This is because if the length of one side is doubled, the area is quadrupled.

As an example, suppose two rectangles are similar, but the edges of the second rectangle are three times longer than the edges of the first rectangle. The area of the first rectangle is 10 square inches. How much more area does the second rectangle have than the first?

To answer this, note that the area of the second rectangle is $3^2 = 9$ times the area of the first rectangle, which is 10 square inches. Therefore, the area of the second rectangle is going to be $9 \times 10 = 90$ square inches. This means it has $90 - 10 = 80$ square inches more area than the first rectangle.

As a second example, suppose *X* and *Y* are similar right triangles. The hypotenuse of *X* is 4 inches. The area of *Y* is $\frac{1}{4}$ the area of *X*. What is the hypotenuse of *Y*?

First, realize the area has changed by a factor of $\frac{1}{4}$. The area changes by a factor that is the *square* of the ratio of changes in lengths, so the ratio of the lengths is the square root of the ratio of areas. That means that the ratio of lengths must be is $\sqrt{\frac{1}{4}} = \frac{1}{2}$, and the hypotenuse of *Y* must be:

$$\frac{1}{2} \times 4 = 2 \text{ inches}$$

Volumes between similar solids change like the cube of the change in the lengths of their edges. Likewise, if the ratio of the volumes between similar solids is known, the ratio between their lengths is known by finding the cube root of the ratio of their volumes.

For example, suppose there are two similar rectangular pyramids *X* and *Y*. The base of *X* is 1 inch by 2 inches, and the volume of *X* is 8 inches. The volume of *Y* is 64 inches. What are the dimensions of the base of *Y*?

To answer this, first find the ratio of the volume of *Y* to the volume of *X*. This will be given by $\frac{64}{8} = 8$. Now the ratio of lengths is the cube root of the ratio of volumes, or $\sqrt[3]{8} = 2$. So, the dimensions of the base of *Y* must be 2 inches by 4 inches.

Problem Solving

The Quantitative section of the GMAT contains two types of questions: problem solving and data sufficiency. While the format and objective of each these types of questions are different, the domains of mathematics addressed in either type of question are the same and include the following:

- Arithmetic
- Elementary algebra
- Common Geometry Concepts
- Word Problems

The format of problem-solving questions will likely be familiar for most test takers. These present a problem that addresses the basic mathematical skills and concepts listed above and offers five possible answer choices. Test takers must solve the problem and then select the single correct answer choice. The problem-solving questions are designed to be at a difficulty level that is on par with what is expected of an eleventh-grade student. Test takers should expect to interpret graphics, demonstrate an understanding of elementary math concepts, reason quantitatively, and solve quantitative problems.

Data Sufficiency

Test takers should expect approximately 14–16 data sufficiency problems in the Quantitative section of the GMAT. For these questions, rather than needing to apply quantitative reasoning methods to explicitly solve the problem, test takers must review the provided information in relation to the posed question and determine if the provided information is sufficient to answer the question.

The data sufficiency questions provide a problem and then two statements labeled 1 and 2. Following this information, five answer choices, which will be identical for all data sufficiency questions, are presented.

Test takers must select the single valid choice for each data sufficiency question from the following five options:

A. Statement (1) ALONE is sufficient, but statement (2) alone is not sufficient to answer the question asked.
B. Statement (2) ALONE is sufficient, but statement (1) alone is not sufficient to answer the question asked.
C. BOTH statements (1) and (2) TOGETHER are sufficient to answer the question asked, but NEITHER statement ALONE is sufficient to answer the question asked.
D. EACH statement ALONE is sufficient to answer the question asked.
E. Statements (1) and (2) TOGETHER are NOT sufficient to answer the question asked, and additional data specific to the problem are needed.

Essentially, test takers must determine if just statement 1 *or* statement 2 alone provide enough information to solve the problem, if both statements are needed to arrive at the answer, if either has enough information (such that statement 1 could be used alone to solve the problem or statement 2 alone, but both aren't necessary), or if the problem cannot be solved even with all of the information put together from both.

The flowchart that follows visually depicts the thinking process required by data sufficiency questions:

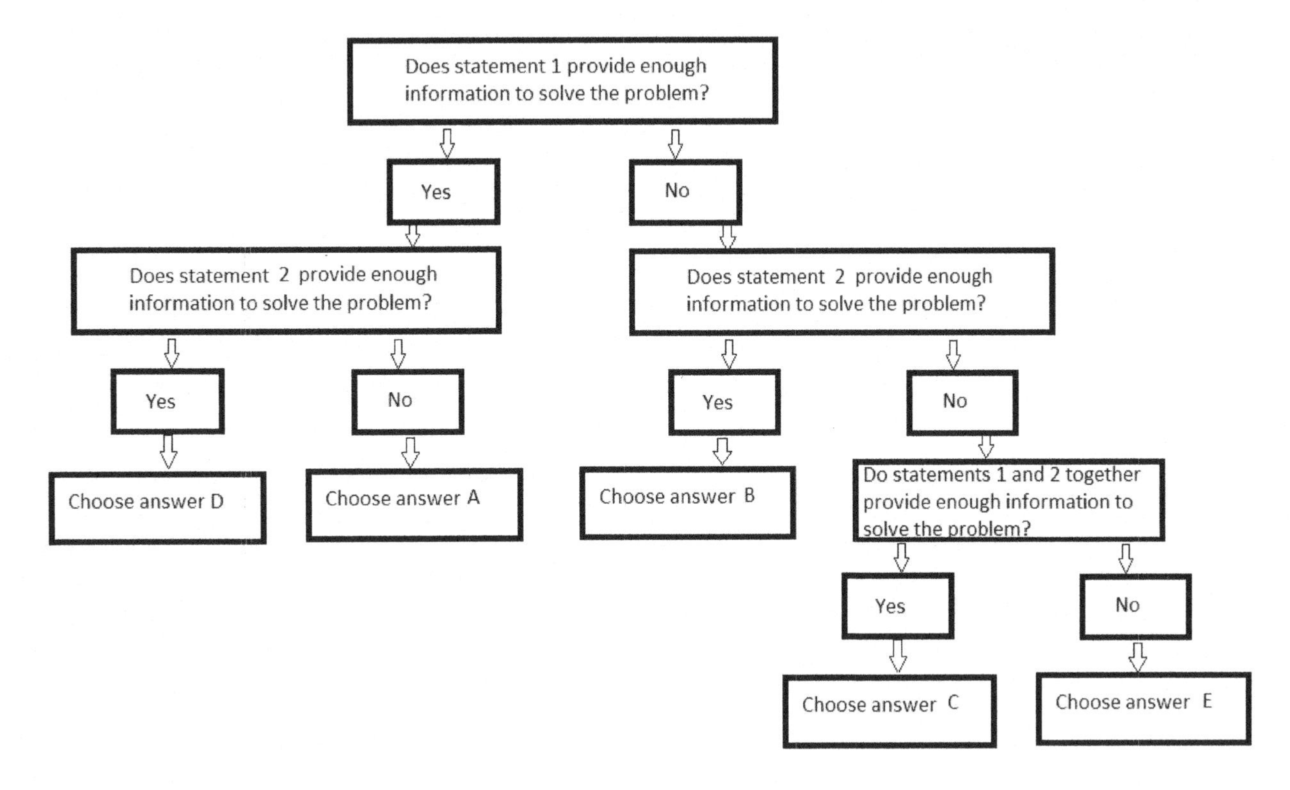

The good news is that test takers should not need to read through each of the five answer choices for every data sufficiency question because they will always be exactly the same and in the same order. By memorizing the answer options, test takers can save some time and bank it to use on actually working through the problems themselves. It is usually time efficient to consider statement 1 in isolation first, then statement 2, and determine if either alone provides information sufficient to solve the problem, or if both are needed. It is important for test takers to take the time to evaluate if either statement can be used in isolation to solve the problem and not just select one of the two because Choice *D* states that statement 1 alone *and* statement 2 alone contain enough information to solve the problem.

Verbal Reasoning

Reading Comprehension

Purpose of a Passage

No matter the genre or format, all authors are writing to persuade, inform, entertain, or express feelings. Often, these purposes are blended, with one dominating the rest. It's useful to learn to recognize the author's intent.

Persuasive writing is used to persuade or convince readers of something. It often contains two elements: the argument and the counterargument. The argument takes a stance on an issue, while the counterargument pokes holes in the opposition's stance. Authors rely on logic, emotion, and writer credibility to persuade readers to agree with them. If readers are opposed to the stance before reading, they are unlikely to adopt that stance. However, those who are undecided or committed to the same stance are more likely to agree with the author.

Informative writing tries to teach or inform. Workplace manuals, instructor lessons, statistical reports, and cookbooks are examples of informative texts. Informative writing is usually based on facts and is often void of emotion and persuasion. Informative texts generally contain statistics, charts, and graphs. Though most informative texts lack a persuasive agenda, readers still must examine the text carefully to determine whether one exists within a given passage.

Stories or narratives are designed to entertain. When you go to the movies, you often want to escape for a few hours, not necessarily to think critically. Entertaining writing is designed to delight and engage the reader. However, sometimes this type of writing can be woven into more serious materials, such as persuasive or informative writing to hook the reader before transitioning into a more scholarly discussion.

Emotional writing works to evoke the reader's feelings, such as anger, euphoria, or sadness. The connection between reader and author is an attempt to cause the reader to share the author's intended emotion or tone. Sometimes in order to make a piece more poignant, the author simply wants readers to feel the same emotions that the author has felt. Other times, the author attempts to persuade or manipulate the reader into adopting his stance. While it's okay to sympathize with the author, be aware of the individual's underlying intent.

Types of Passages

Writing can be classified under four passage types: narrative, expository, descriptive (sometimes called technical), and persuasive. Though these types are not mutually exclusive, one form tends to dominate the rest. By recognizing the *type* of passage you're reading, you gain insight into *how* you should read. If you're reading a narrative, you can assume the author intends to entertain, which means you may skim the text without losing meaning. A technical document might require a close read because skimming the passage might cause the reader to miss salient details.

1. **Narrative** writing, at its core, is the art of storytelling. For a narrative to exist, certain elements must be present. First, it must have characters. While many characters are human, characters could be defined as anything that thinks, acts, and talks like a human. For example, many recent movies, such as

Lord of the Rings and *The Chronicles of Narnia*, include animals, fantastical creatures, and even trees that behave like humans. Second, it must have a plot or sequence of events. Typically, those events follow a standard plot diagram, but recent trends start *in medias res* or in the middle (near the climax). In this instance, foreshadowing and flashbacks often fill in plot details. Finally, along with characters and a plot, there must also be conflict. Conflict is usually divided into two types: internal and external. Internal conflict indicates the character is in turmoil and is presented through the character's thoughts. External conflicts are visible. Types of external conflict include a person versus nature, another person, or society.

2. **Expository** writing is detached and to the point. Since expository writing is designed to instruct or inform, it usually involves directions and steps written in second person ("you" voice) and lacks any persuasive or narrative elements. Sequence words such as *first*, *second*, and *third*, or *in the first place*, *secondly*, and *lastly* are often given to add fluency and cohesion. Common examples of expository writing include instructor's lessons, cookbook recipes, and repair manuals.

3. Due to its empirical nature, **technical** writing is filled with steps, charts, graphs, data, and statistics. The goal of technical writing is to advance understanding in a field through the scientific method. Experts such as teachers, doctors, or mechanics use words unique to the profession in which they operate. These words, which often incorporate acronyms, are called **jargon**. Technical writing is a type of expository writing but is not meant to be understood by the general public. Instead, technical writers assume readers have received a formal education in a particular field of study and need no explanation as to what the jargon means. Imagine a doctor trying to understand a diagnostic reading for a car or a mechanic trying to interpret lab results. Only professionals with proper training will fully comprehend the text.

4. **Persuasive** writing is designed to change opinions and attitudes. The topic, stance, and arguments are found in the thesis, positioned near the end of the introduction. Later supporting paragraphs offer relevant quotations, paraphrases, and summaries from primary or secondary sources, which are then interpreted, analyzed, and evaluated. The goal of persuasive writers is not to stack quotes but to develop original ideas by using sources as a starting point. Good persuasive writing makes powerful arguments with valid sources and thoughtful analysis. Poor persuasive writing is riddled with bias and logical fallacies. Sometimes logical and illogical arguments are sandwiched together in the same piece. Therefore, readers should display skepticism when reading persuasive arguments.

Text Structure

Depending on what the author is attempting to accomplish, certain formats or text structures work better than others. For example, a sequence structure might work for narration but not for identifying similarities and differences between concepts. Similarly, a comparison-contrast structure is not useful for narration. It's the author's job to put the right information in the correct format.

Readers should be familiar with the five main literary structures:

1. **Sequence** structure (sometimes referred to as the order structure) is when the order of events proceed in a predictable order. In many cases, this means the text goes through the plot elements: exposition, rising action, climax, falling action, and resolution. Readers are introduced to characters, setting, and conflict in the exposition. In the rising action, there's an increase in tension and suspense. The climax is the height of tension and the point of no return. Tension decreases during the falling

action. In the resolution, any conflicts presented in the exposition are solved, and the story concludes. An informative text that is structured sequentially will often go in order from one step to the next.

2. In the **problem-solution** structure, authors identify a potential problem and suggest a solution. This form of writing is usually divided into two paragraphs and can be found in informational texts. For example, cell phone, cable, and satellite providers use this structure in manuals to help customers troubleshoot or identify problems with services or products.

3. When authors want to discuss similarities and differences between separate concepts, they arrange thoughts in a **comparison-contrast** paragraph structure. Venn diagrams are an effective graphic organizer for comparison-contrast structures because they feature two overlapping circles that can be used to organize similarities and differences. A comparison-contrast essay organizes one paragraph based on similarities and another based on differences. A comparison-contrast essay can also be arranged with the similarities and differences of individual traits addressed within individual paragraphs. Words such as *however*, *but*, and *nevertheless* help signal a contrast in ideas.

4. **Descriptive** writing structure is designed to appeal to your senses. Much like an artist who constructs a painting, good descriptive writing builds an image in the reader's mind by appealing to the five senses: sight, hearing, taste, touch, and smell. However, overly descriptive writing can become tedious; whereas sparse descriptions can make settings and characters seem flat. Good authors strike a balance by applying descriptions only to passages, characters, and settings that are integral to the plot.

5. Passages that use the **cause and effect** structure are simply asking *why* by demonstrating some type of connection between ideas. Words such as *if, since, because, then, or consequently* indicate a cause-and-effect relationship. By switching the order of a complex sentence, the writer can rearrange the emphasis on different clauses. Saying, *If Sheryl is late, we'll miss the dance*, is different from saying *We'll miss the dance if Sheryl is late*. One emphasizes Sheryl's tardiness while the other emphasizes missing the dance. Paragraphs can also be arranged in a cause and effect format. Since the format—before and after—is sequential, it is useful when authors wish to discuss the impact of choices. Researchers often apply this paragraph structure to the scientific method.

Point of View

Point of view is an important writing device to consider. In fiction writing, point of view refers to who tells the story or from whose perspective readers are observing the story. In non-fiction writing, the **point of view** refers to whether the author refers to himself/herself, their readers, or chooses not to mention either. Whether fiction or nonfiction, the author will carefully consider the impact the perspective will have on the purpose and main point of the writing.

- **First-person point of view**: The story is told from the writer's perspective. In fiction, this would mean that the main character is also the narrator. First-person point of view is easily recognized by the use of personal pronouns such as *I*, *me*, *we*, *us*, *our*, *my*, and *myself*.

- **Third-person point of view**: In a more formal essay, this would be an appropriate perspective because the focus should be on the subject matter, not the writer or the reader. Third-person point of view is recognized by the use of the pronouns *he*, *she*, *they*, and *it*. In fiction writing, third-person point of view has a few variations.

- **Third-person limited** point of view refers to a story told by a narrator who has access to the thoughts and feelings of just one character.

- In **third-person omniscient** point of view, the narrator has access to the thoughts and feelings of all the characters.
- In **third-person objective** point of view, the narrator is like a fly on the wall and can see and hear what the characters do and say but does not have access to their thoughts and feelings.

- **Second-person point of view**: This point of view isn't commonly used in fiction or nonfiction writing because it directly addresses the reader using the pronouns *you*, *your*, and *yourself*. Second-person perspective is more appropriate in direct communication, such as business letters or emails.

Point of View	Pronouns Used
First person	I, me, we, us, our, my, myself
Second person	You, your, yourself
Third person	He, she, it, they

Main Ideas and Supporting Details

Topics and main ideas are critical parts of writing. The **topic** is the subject matter of the piece. An example of a topic would be *global warming*.

The main idea is what the writer wants to say about that topic. A writer may make the point that global warming is a growing problem that must be addressed in order to save the planet. Therefore, the topic is global warming, and the main idea is that it's *a serious problem needing to be addressed*. The topic can be expressed in a word or two, but the main idea should be a complete thought.

An author will likely identify the topic immediately within the title or the first sentence of a passage. The main idea is usually presented in the introduction. In a single passage, the main idea may be identified in the first or last sentence, but it will most likely be directly stated and easily recognized by the reader. Because it is not always stated immediately in a passage, it's important to carefully read the entire passage to identify the main idea.

The main idea should not be confused with the thesis statement. A **thesis statement** is a clear statement of the writer's specific stance and can often be found in the introduction of a non-fiction piece. The thesis is a specific sentence (or two) that offers the direction and focus of the discussion.

In order to illustrate the main idea, a writer will use **supporting details**, the details that provide evidence or examples to help make a point. Supporting details often appear in the form of quotations, paraphrasing, or analysis. Authors should connect details and analysis to the main point.

For example, in the example of global warming, where the author's main idea is to show the seriousness of this growing problem and the need for change, the use of supporting details in this piece would be critical in effectively making that point. Supporting details used here might include statistics on an increase in global temperatures and studies showing the impact of global warming on the planet. The author could also include projections for future climate change in order to illustrate potential lasting effects of global warming.

It's important to evaluate the author's supporting details to be sure that they are credible, provide evidence of the author's point, and directly support the main idea. Though shocking statistics grab readers' attention, their use could be ineffective information in the piece. Details like this are crucial to understanding the passage and evaluating how well the author presents their argument and evidence.

Also remember that when most authors write, they want to make a point or send a message. This point or message of a text is known as the theme. Authors may state themes explicitly, like in *Aesop's Fables*. More often, especially in modern literature, readers must infer the theme based on text details. Usually after carefully reading and analyzing an entire text, the reader can identify the theme. Typically, the longer the piece, the more themes you will encounter, though often one theme dominates the rest, as evidenced by the author's purposeful revisiting of it throughout the passage.

Evaluating a Passage

Determining conclusions requires being an active reader, as a reader must make a prediction and analyze facts to identify a conclusion. There are a few ways to determine a logical conclusion, but careful reading is the most important. It's helpful to read a passage a few times, noting details that seem important to the piece. A reader should also identify key words in a passage to determine the logical conclusion or determination that flows from the information presented.

Textual evidence within the details helps readers draw a conclusion about a passage. **Textual evidence** refers to information—facts and examples that support the main point. Textual evidence will likely come from outside sources and can be in the form of quoted or paraphrased material. In order to draw a conclusion from evidence, it's important to examine the credibility and validity of that evidence as well as how (and if) it relates to the main idea.

If an author presents a differing opinion or a **counter-argument** in order to refute it, the reader should consider how and why the information is being presented. It is meant to strengthen the original argument and shouldn't be confused with the author's intended conclusion, but it should also be considered in the reader's final evaluation.

Sometimes, authors explicitly state the conclusion they want readers to understand. Alternatively, a conclusion may not be directly stated. In that case, readers must rely on the implications to form a logical conclusion:

> On the way to the bus stop, Michael realized his homework wasn't in his backpack. He ran back to the house to get it and made it back to the bus just in time.

In this example, though it's never explicitly stated, it can be inferred that Michael is a student on his way to school in the morning. When forming a conclusion from implied information, it's important to read the text carefully to find several pieces of evidence to support the conclusion.

Summarizing is an effective way to draw a conclusion from a passage. A summary is a shortened version of the original text, written by the reader in their own words. Focusing on the main points of the original text and including only the relevant details can help readers reach a conclusion. It's important to retain the original meaning of the passage.

Like summarizing, **paraphrasing** can also help a reader fully understand different parts of a text. Paraphrasing calls for the reader to take a small part of the passage and list or describe its main points. Paraphrasing is more than rewording the original passage, though. It should be written in the reader's own words, while still retaining the meaning of the original source. This will indicate an understanding of the original source, yet still help the reader expand on their interpretation.

Readers should pay attention to the **sequence**, or the order in which details are laid out in the text, as this can be important to understanding its meaning as a whole. Writers will often use transitional words to help the reader understand the order of events and to stay on track. Words like *next, then, after*, and *finally* show that the order of events is important to the author. In some cases, the author omits these transitional words, and the sequence is implied. Authors may even purposely present the information out of order to make an impact or have an effect on the reader. An example might be when a narrative writer uses **flashback** to reveal information.

Responding to a Passage

There are a few ways for readers to engage actively with the text, such as making inferences and predictions. An **inference** refers to a point that is implied (as opposed to directly-stated) by the evidence presented:

> Bradley packed up all of the items from his desk in a box and said goodbye to his coworkers for the last time.

From this sentence, though it is not directly stated, readers can infer that Bradley is leaving his job. It's necessary to use inference in order to draw conclusions about the meaning of a passage. Authors make implications through character dialogue, thoughts, effects on others, actions, and looks. Like in life, readers must assemble all the clues to form a complete picture.

When making an inference about a passage, it's important to rely only on the information that is provided in the text itself. This helps readers ensure that their conclusions are valid.

Readers will also find themselves making predictions when reading a passage or paragraph. **Predictions** are guesses about what's going to happen next. Readers can use prior knowledge to help make accurate predictions. Prior knowledge is best utilized when readers make links between the current text, previously read texts, and life experiences. Some texts use suspense and foreshadowing to captivate readers:

A cat darted across the street just as the car came careening around the curve.

One unfortunate prediction might be that the car will hit the cat. Of course, predictions aren't always accurate, so it's important to read carefully to the end of the text to determine the accuracy of predictions.

Critical Reasoning

Critical Thinking Skills

It's important to read any piece of writing critically. The goal is to discover the point and purpose of what the author is writing about through analysis. It's also crucial to establish the point or stance the author has taken on the topic of the piece. After determining the author's perspective, readers can then more effectively develop their own viewpoints on the subject of the piece.

It is important to distinguish between fact and opinion when reading a piece of writing. A **fact** is information that can be proven true. If information can be disproved, it is not a fact. For example, water freezes at or below thirty-two degrees Fahrenheit. An argument stating that water freezes at seventy degrees Fahrenheit cannot be supported by data and is therefore not a fact. Facts tend to be associated with science, mathematics, and statistics. Opinions are information open to debate. Opinions are often tied to subjective concepts like equality, morals, and rights. They can also be controversial.

Authors often use words like *think, feel, believe,* or *in my opinion* when expressing opinion, but these words won't always appear in an opinion piece, especially if it is formally written. An author's opinion may be backed up by facts, which gives it more credibility, but that opinion should not be taken as fact. A critical reader should be wary of an author's opinion, especially if it is only supported by other opinions.

Fact	**Opinion**
There are 9 innings in a game of baseball.	Baseball games run too long.
James Garfield was assassinated on July 2, 1881.	James Garfield was a good president.
McDonalds has stores in 118 countries.	McDonalds has the best hamburgers.

Critical readers examine the facts used to support an author's argument. They check the facts against other sources to be sure those facts are correct. They also check the validity of the sources used to be sure those sources are credible, academic, and/or peer-reviewed. Consider that when an author uses another person's opinion to support their argument, even if it is an expert's opinion, it is still only an opinion and should not be taken as fact. A strong argument uses valid, measurable facts to support ideas. Even then, the reader may disagree with the argument as it may be rooted in their personal beliefs.

An authoritative argument may use the facts to sway the reader. Because of this, a writer may choose to only use the information and expert opinion that supports their viewpoint.

If the argument is that wind energy is the best solution, the author will use facts that support this idea. That same author may leave out relevant facts on solar energy. The way the author uses facts can influence the reader, so it's important to consider the facts being used, how those facts are being presented, and what information might be left out.

Critical readers should also look for errors in the argument such as logical fallacies and bias. A **logical fallacy** is a flaw in the logic used to make the argument. Logical fallacies include slippery slope, straw man, and begging the question. Authors can also reflect **bias** if they ignore an opposing viewpoint or present their side in an unbalanced way. A strong argument considers the opposition and finds a way to refute it. Critical readers should look for an unfair or one-sided presentation of the argument and be skeptical, as a bias may be present. Even if this bias is unintentional, if it exists in the writing, the reader should be wary of the validity of the argument.

Readers should also look for the use of **stereotypes**, which refer to specific groups. Stereotypes are often negative connotations about a person or place and should always be avoided. When a critical reader finds stereotypes in a piece of writing, they should immediately be critical of the argument and consider the validity of anything the author presents. Stereotypes reveal a flaw in the writer's thinking and may suggest a lack of knowledge or understanding about the subject.

Tone

Tone refers to the writer's attitude toward the subject matter. Tone is usually explained in terms of a work of fiction. For example, the tone conveys how the writer feels about their characters and the situations in which they're involved. Nonfiction writing is sometimes thought to have no tone at all; however, this is incorrect.

A lot of nonfiction writing has a neutral tone, which is an important tone for the writer to take. A neutral tone demonstrates that the writer is presenting a topic impartially and letting the information speak for itself. On the other hand, nonfiction writing can be just as effective and appropriate if the tone isn't neutral. For instance, take this example involving seat belts:

> Seat belts save more lives than any other automobile safety feature. Many studies show that airbags save lives as well; however, not all cars have airbags. For instance, some older cars don't. Furthermore, air bags aren't entirely reliable. For example, studies show that in 15% of accidents airbags don't deploy as designed, but, on the other hand, seat belt malfunctions are extremely rare. The number of highway fatalities has plummeted since laws requiring seat belt usage were enacted.

In this passage, the writer mostly chooses to retain a neutral tone when presenting information. If the writer would instead include their own personal experience of losing a friend or family member in a car accident, the tone would change dramatically. The tone would no longer be neutral and would show that the writer has a personal stake in the content, allowing them to interpret the information in a different way. When analyzing tone, consider what the writer is trying to achieve in the text and how they *create* the tone using style.

Evaluation of Sources

Identifying Relevant Information During Research

Relevant information is that which is pertinent to the topic at hand. Particularly when doing research online, it is easy for students to get overwhelmed with the wealth of information available to them. Before conducting research, then, students need to begin with a clear idea of the question they want to answer.

For example, a student may be interested in learning more about marriage practices in Jane Austen's England. If that student types "marriage" into a search engine, he or she will have to sift through thousands of unrelated sites before finding anything related to that topic. Narrowing down search parameters can aid in locating relevant information.

When using a book, students can consult the table of contents, glossary, or index to discover whether the book contains relevant information before using it as a resource. If the student finds a hefty volume on Jane Austen, he or she can flip to the index in the back, look for the word marriage, and find out how many page references are listed in the book. If there are few or no references to the subject, it is probably not a relevant or useful source.

In evaluating research articles, students may also consult the title, abstract, and keywords before reading the article in its entirety. Referring to the date of publication will also determine whether the research contains up-to-date discoveries, theories, and ideas about the subject, or whether it is outdated.

Evaluating the Credibility of a Print or Digital Source

There are several additional criteria that need to be examined before using a source for a research topic.

The following questions will help determine whether a source is credible:

- Author
  - Who is he or she?
  - Does he or she have the appropriate credentials—e.g., M.D, PhD?
  - Is this person authorized to write on the matter through their job or personal experiences?
  - Is he or she affiliated with any known credible individuals or organizations?
  - Has he or she written anything else?
- Publisher
  - Who published/produced the work? Is it a well-known journal, like National Geographic, or a tabloid, like The National Enquirer?
  - Is the publisher from a scholarly, commercial, or government association?
  - Do they publish works related to specific fields?
  - Have they published other works?
  - If a digital source, what kind of website hosts the text? Does it end in .edu, .org, or .com?
- Bias
  - Is the writing objective? Does it contain any loaded or emotional language?
  - Does the publisher/producer have a known bias, such as Fox News or CNN?
  - Does the work include diverse opinions or perspectives?
  - Does the author have any known bias—e.g., Michael Moore, Bill O'Reilly, or the Pope? Is he or she affiliated with any organizations or individuals that may have a known bias—e.g., Citizens United or the National Rifle Association?
  - Does the magazine, book, journal, or website contain any advertising?
- References
  - Are there any references?
  - Are the references credible? Do they follow the same criteria as stated above?
  - Are the references from a related field?
- Accuracy/reliability
  - Has the article, book, or digital source been peer reviewed?
  - Are all of the conclusions, supporting details, or ideas backed with published evidence?
  - If a digital source, is it free of grammatical errors, poor spelling, and improper English?
  - Do other published individuals have similar findings?
- Coverage
  - Are the topic and related material both successfully addressed?
  - Does the work add new information or theories to those of their sources?
  - Is the target audience appropriate for the intended purpose?

Sentence Correction

Types of Sentences

There isn't an overabundance of absolutes in grammar, but here is one: every sentence in the English language falls into one of four categories.

- Declarative: a simple statement that ends with a period

    The price of milk per gallon is the same as the price of gasoline.

- Imperative: a command, instruction, or request that ends with a period

    Buy milk when you stop to fill up your car with gas.

- Interrogative: a question that ends with a question mark

    Will you buy the milk?

- Exclamatory: a statement or command that expresses emotions like anger, urgency, or surprise and ends with an exclamation mark

    Buy the milk now!

Declarative sentences are the most common type, probably because they are comprised of the most general content, without any of the bells and whistles that the other three types contain. They are, simply, declarations or statements of any degree of seriousness, importance, or information.

Imperative sentences often seem to be missing a subject. The subject is there, though; it is just not visible or audible because it is implied. Look at the imperative example sentence.

Buy the milk when you fill up your car with gas.

You is the implied subject, the one to whom the command is issued. This is sometimes called *the understood you* because it is understood that *you* is the subject of the sentence.

Interrogative sentences—those that ask questions—are defined as such from the idea of the word *interrogation*, the action of questions being asked of suspects by investigators. Although that is serious business, interrogative sentences apply to all kinds of questions.

To exclaim is at the root of **exclamatory** sentences. These are made with strong emotions behind them. The only technical difference between a declarative or imperative sentence and an exclamatory one is the exclamation mark at the end. The example declarative and imperative sentences can both become an exclamatory one simply by putting an exclamation mark at the end of the sentences.

The price of milk per gallon is the same as the price of gasoline!

Buy milk when you stop to fill up your car with gas!

After all, someone might be really excited by the price of gas or milk, or they could be mad at the person that will be buying the milk! However, as stated before, exclamation marks in abundance defeat their own purpose! After a while, they begin to cause fatigue! When used only for their intended purpose, they can have their expected and desired effect.

Parts of Speech

Nouns

A **noun** is a person, place, thing, or idea. All nouns fit into one of two types: common or proper.

A **common noun** is a word that identifies any of a class of people, places, or things. Examples include numbers, objects, animals, feelings, concepts, qualities, and actions. *A, an,* or *the* usually precedes the common noun. These parts of speech are called **articles**. Here are some examples of sentences using nouns preceded by articles.

> A building is under construction.
>
> The girl would like to move to the city.

A **proper noun** is used for the specific name of an individual person, place, or organization. The first letter in a proper noun is capitalized. "My name is *Mary*." "I work for *Walmart*."

Nouns sometimes serve as adjectives (which themselves describe nouns), such as "hockey player" and "state government."

An **abstract noun** is an idea, state, or quality. It is something that can't be touched, such as happiness, courage, evil, or humor.

A **concrete noun** is something that can be experienced through the senses (touch, taste, hear, smell, see). Examples of concrete nouns are birds, skateboard, pie, and car.

A **collective noun** refers to a collection of people, places, or things that act as one. Examples of collective nouns are as follows: team, class, jury, family, audience, and flock.

Pronouns

A word used in place of a noun is known as a **pronoun**. Pronouns are words like *I, mine, hers,* and *us*.

Pronouns can be split into different classifications (see below) which make them easier to learn; however, it's not important to memorize the classifications.

- Personal pronouns: refer to people
  - First person: we, I, our, mine
  - Second person: you, yours
  - Third person: he, them
- Possessive pronouns: demonstrate ownership (mine, his, hers, its, ours, theirs, yours)
- Interrogative pronouns: ask questions (what, which, who, whom, whose)
- Relative pronouns: include the five interrogative pronouns and others that are relative (whoever, whomever, that, when, where)
- Demonstrative pronouns: replace something specific (this, that, those, these)

- Reciprocal pronouns: indicate something was done or given in return (each other, one another)
- **Indefinite pronouns:** have a nonspecific status (anybody, whoever, someone, everybody, somebody)
    - **Indefinite pronouns** such as *anybody, whoever, someone, everybody*, and *somebody* command a singular verb form, but others such as *all, none,* and *some* could require a singular or plural verb form.

Antecedents

An **antecedent** is the noun to which a pronoun refers; it needs to be written or spoken before the pronoun is used. For many pronouns, antecedents are imperative for clarity. In particular, many of the personal, possessive, and demonstrative pronouns need antecedents. Otherwise, it would be unclear who or what someone is referring to when they use a pronoun like *he* or *this*.

Pronoun reference means that the pronoun should refer clearly to one, clear, unmistakable noun (the antecedent).

Pronoun-antecedent agreement refers to the need for the antecedent and the corresponding pronoun to agree in gender, person, and number. Here are some examples:

> The *kidneys* (plural antecedent) are part of the urinary system. *They* (plural pronoun) serve several roles.
>
> The kidneys are part of the *urinary system* (singular antecedent). *It* (singular pronoun) is also known as the renal system.

Pronoun Cases

The subjective pronouns —*I, you, he/she/it, we, they,* and *who*—are the subjects of the sentence.

> Example: *They* have a new house.

The objective pronouns—*me, you* (*singular*), *him/her, us, them,* and *whom*—are used when something is being done for or given to someone; they are objects of the action.

> Example: The teacher has an apple for *us*.

The possessive pronouns—*mine, my, your, yours, his, hers, its, their, theirs, our,* and *ours*—are used to denote that something (or someone) belongs to someone (or something).

> Example: It's their chocolate cake.
>
> Even Better Example: It's my chocolate cake!

One of the greatest challenges and worst abuses of pronouns concerns *who* and *whom*. Just knowing the following rule can eliminate confusion. *Who* is a subjective-case pronoun used only as a subject or subject complement. *Whom* is only objective-case and, therefore, the object of the verb or preposition.

> *Who* is going to the concert?
>
> You are going to the concert with *whom*?

Hint: When using *who* or *whom*, think of whether someone would say *he* or *him*. If the answer is *he*, use *who*. If the answer is *him*, use *whom*. This trick is easy to remember because *he* and *who* both end in vowels, and *him* and *whom* both end in the letter *M*.

Verbs

A **verb** is the part of speech that describes an action, state of being, or occurrence.

A verb forms the main part of a predicate of a sentence. This means that the verb explains what the noun (which will be discussed shortly) is doing. A simple example is *time flies*. The verb *flies* explains what the action of the noun, *time*, is doing. This example is a **main** verb.

Helping (auxiliary) verbs are words like *have, do, be, can, may, should, must,* and *will.* "I *should* go to the store." Helping verbs assist main verbs in expressing tense, ability, possibility, permission, or obligation.

Particles are minor function words like *not, in, out, up,* or *down* that become part of the verb itself. "I might *not*."

Participles are words formed from verbs that are often used to modify a noun, noun phrase, verb, or verb phrase.

> The *running* teenager collided with the cyclist.

Participles can also create compound verb forms.

> He is *speaking*.

Verbs have five basic forms: the **base** form, the ***-s*** form, the ***-ing*** form, the **past** form, and the **past participle** form.

The past forms are either **regular** (*love/loved; hate/hated*) or **irregular** because they don't end by adding the common past tense suffix "-ed" (*go/went; fall/fell; set/set*).

Verb Forms

Shifting verb forms entails **conjugation**, which is used to indicate tense, voice, or mood.

Verb tense is used to show when the action in the sentence took place. There are several different verb tenses, and it is important to know how and when to use them. Some verb tenses can be achieved by changing the form of the verb, while others require the use of helping verbs (e.g., *is, was,* or *has*).

Present tense shows the action is happening currently or is ongoing:

> I walk to work every morning.
>
> She is stressed about the deadline.

Past tense shows that the action happened in the past or that the state of being is in the past:

> I walked to work yesterday morning.
>
> She was stressed about the deadline.

Future tense shows that the action will happen in the future or is a future state of being:

I will walk to work tomorrow morning.

She will be stressed about the deadline.

Present perfect tense shows action that began in the past, but continues into the present:

I have walked to work all week.

She has been stressed about the deadline.

Past perfect tense shows an action was finished before another took place:

I had walked all week until I sprained my ankle.

She had been stressed about the deadline until we talked about it.

Future perfect tense shows an action that will be completed at some point in the future:

By the time the bus arrives, I will have walked to work already.

Voice

Verbs can be in the active or passive voice. When the subject completes the action, the verb is in **active voice**. When the subject receives the action of the sentence, the verb is in **passive voice**.

Active: Jamie ate the ice cream.

Passive: The ice cream was eaten by Jamie.

In active voice, the subject (*Jamie*) is the "do-er" of the action (*ate*). In passive voice, the subject *ice cream* receives the action of being eaten.

While passive voice can add variety to writing, active voice is the generally preferred sentence structure.

Mood

Mood is used to show the speaker's feelings about the subject matter. In English, there is indicative mood, imperative mood, and subjective mood.

Indicative mood is used to state facts, ask questions, or state opinions:

Bob will make the trip next week.

When can Bob make the trip?

Imperative mood is used to state a command or make a request:

Wait in the lobby.

Please call me next week.

Subjunctive mood is used to express a wish, an opinion, or a hope that is contrary to fact:

If I were in charge, none of this would have happened.

Allison wished she could take the exam over again when she saw her score.

Adjectives

Adjectives are words used to modify nouns and pronouns. They can be used alone or in a series and are used to further define or describe the nouns they modify.

Mark made us a delicious, four-course meal.

The words *delicious* and *four-course* are adjectives that describe the kind of meal Mark made.

Articles are also considered adjectives because they help to describe nouns. Articles can be general or specific. The three articles in English are: a, an, and the.

Indefinite articles *(a, an)* are used to refer to nonspecific nouns. The article *a* proceeds words beginning with consonant sounds, and the article *an* proceeds words beginning with vowel sounds.

A car drove by our house.

An alligator was loose at the zoo.

He has always wanted a ukulele. (The first *u* makes a *y* sound.)

Note that *a* and *an* should only proceed nonspecific nouns that are also singular. If a nonspecific noun is plural, it does not need a preceding article.

Alligators were loose at the zoo.

The **definite article** *(the)* is used to refer to specific nouns:

The car pulled into our driveway.

Note that *the* should proceed all specific nouns regardless of whether they are singular or plural.

The cars pulled into our driveway.

Comparative adjectives are used to compare nouns. When they are used in this way, they take on positive, comparative, or superlative form.

The **positive** form is the normal form of the adjective:
Alicia is tall.

The **comparative** form shows a comparison between two things:

Alicia is taller than Maria.

Superlative form shows comparison between more than two things:

Alicia is the tallest girl in her class.

Usually, the comparative and superlative can be made by adding *–er* and *–est* to the positive form, but some verbs call for the helping verbs *more* or *most.* Other exceptions to the rule include adjectives like *bad*, which uses the comparative *worse* and the superlative *worst*.

An adjective phrase is not a bunch of adjectives strung together, but a group of words that describes a noun or pronoun and, thus, functions as an adjective. Very happy is an adjective phrase; so are way too hungry and passionate about traveling.

Adverbs

Adverbs have more functions than adjectives because they modify or qualify verbs, adjectives, or other adverbs as well as word groups that express a relation of place, time, circumstance, or cause. Therefore, adverbs answer any of the following questions: *How, when, where, why, in what way, how often, how much, in what condition,* and/or *to what degree. How good looking is he? He is very handsome.*

Here are some examples of adverbs for different situations:

- how: quickly
- when: daily
- where: there
- in what way: easily
- how often: often
- how much: much
- in what condition: badly
- what degree: hardly

As one can see, for some reason, many adverbs end in *-ly*.

Adverbs do things like emphasize (*really, simply,* and *so*), amplify (*heartily, completely,* and *positively*), and tone down (*almost, somewhat,* and *mildly*).

Adverbs also come in phrases.

> The dog ran as *though his life depended on it.*

Prepositions

Prepositions are connecting words and, while there are only about 150 of them, they are used more often than any other individual groups of words. They describe relationships between other words. They are placed before a noun or pronoun, forming a phrase that modifies another word in the sentence. **Prepositional phrases** begin with a preposition and end with a noun or pronoun, the **object of the preposition**. *A pristine lake is near the store and behind the bank.*

Some commonly used prepositions are *about, after, anti, around, as, at, behind, beside, by, for, from, in, into, of, off, on, to,* and *with.*

Complex prepositions, which also come before a noun or pronoun, consist of two or three words such as *according to, in regards to,* and *because of.*

Conjunctions

Conjunctions are vital words that connect words, phrases, thoughts, and ideas. Conjunctions show relationships between components. There are two types:

Coordinating conjunctions are the primary class of conjunctions placed between words, phrases, clauses, and sentences that are of equal grammatical rank; the coordinating conjunctions are *for*, *and*, *nor*, *but*, *or*, *yet*, and *so*. A useful memorization trick is to remember that all the first letters of these conjunctions collectively spell the word FANBOYS.

I need to go shopping, *but* I must be careful to leave enough money in the bank.

She wore a black, red, *and* white shirt.

Subordinating conjunctions are the secondary class of conjunctions. They connect two unequal parts, one **main** (or **independent**) and the other **subordinate** (or **dependent**). I must go to the store *even though* I do not have enough money in the bank.

Because I read the review, I do not want to go to the movie.

Notice that the presence of subordinating conjunctions makes clauses dependent. *I read the review* is an independent clause, but *because* makes the clause dependent. Thus, it needs an independent clause to complete the sentence.

Interjections

Interjections are words used to express emotion. Examples include *wow*, *ouch*, and *hooray*. Interjections are often separate from sentences; in those cases, the interjection is directly followed by an exclamation point. In other cases, the interjection is included in a sentence and followed by a comma. The punctuation plays a big role in the intensity of the emotion that the interjection is expressing. Using a comma or semicolon indicates less excitement than using an exclamation mark.

Capitalization Rules

Here's a non-exhaustive list of things that should be capitalized.

- The first word of every sentence
- The first word of every line of poetry
- The first letter of proper nouns (World War II)
- Holidays (Valentine's Day)
- The days of the week and months of the year (Tuesday, March)
- The first word, last word, and all major words in the titles of books, movies, songs, and other creative works (In the novel, *To Kill a Mockingbird*, note that *a* is lowercase since it's not a major word, but *to* is capitalized since it's the first word of the title.)
- Titles when preceding a proper noun (President Roberto Gonzales, Aunt Judy)

When simply using a word such as president or secretary, though, the word is not capitalized.

Officers of the new business must include a *president* and *treasurer*.

Seasons—spring, fall, etc.—are not capitalized.

North, *south*, *east*, and *west* are capitalized when referring to regions but are not when being used for directions. In general, if it's preceded by *the* it should be capitalized.

I'm from the South.

I drove south.

End Punctuation

Periods (.) are used to end a sentence that is a statement (**declarative**) or a command (**imperative**). They should not be used in a sentence that asks a question or is an exclamation. Periods are also used in abbreviations, which are shortened versions of words.

- Declarative: The boys refused to go to sleep.
- Imperative: Walk down to the bus stop.
- Abbreviations: Joan Roberts, M.D., Apple Inc., Mrs. Adamson
- If a sentence ends with an abbreviation, it is inappropriate to use two periods. It should end with a single period after the abbreviation.

The chef gathered the ingredients for the pie, which included apples, flour, sugar, etc.

Question marks (?) are used with direct questions (**interrogative**). An **indirect question** can use a period:

Interrogative: When does the next bus arrive?

Indirect Question: I wonder when the next bus arrives.

An **exclamation point** (!) is used to show strong emotion or can be used as an **interjection**. This punctuation should be used sparingly in formal writing situations.

What an amazing shot!

Whoa!

Commas

A **comma** (,) is the punctuation mark that signifies a pause—breath—between parts of a sentence. It denotes a break of flow. Proper comma usage helps readers understand the writer's intended emphasis of ideas.

In a complex sentence—one that contains a subordinate (dependent) clause or clauses—the use of a comma is dictated by where the subordinate clause is located. If the subordinate clause is located before the main clause, a comma is needed between the two clauses.

I will not pay for the steak, *because I don't have that much money*.

Generally, if the subordinate clause is placed after the main clause, no punctuation is needed.

I did well on my exam because I studied two hours the night before.

Notice how the last clause is dependent because it requires the earlier independent clauses to make sense.

Use a comma on both sides of an interrupting phrase.

> I will pay for the ice cream, *chocolate and vanilla*, and then will eat it all myself.

The words forming the phrase in italics are nonessential (extra) information. To determine if a phrase is nonessential, try reading the sentence without the phrase and see if it's still coherent.

A comma is not necessary in this next sentence because no interruption—nonessential or extra information—has occurred. Read sentences aloud when uncertain.

> I will pay for his chocolate and vanilla ice cream and then will eat it all myself.

If the nonessential phrase comes at the beginning of a sentence, a comma should only go at the end of the phrase. If the phrase comes at the end of a sentence, a comma should only go at the beginning of the phrase.

Other types of interruptions include the following:

- Interjections: Oh no, I am not going.
- Abbreviations: Barry Potter, M.D., specializes in heart disorders.
- Direct addresses: Yes, Claudia, I am tired and going to bed.
- Parenthetical phrases: His wife, lovely as she was, was not helpful.
- Transitional phrases: Also, it is not possible.

The second comma in the following sentence is called an Oxford comma.

> I will pay for ice cream, syrup, and pop.

It is a comma used after the second-to-last item in a series of three or more items. It comes before the word *or* or *and*. Not everyone uses the Oxford comma; it is optional, but many believe it is needed. The comma functions as a tool to reduce confusion in writing. So, if omitting the Oxford comma would cause confusion, then it's best to include it.

Commas are used in math to mark the place of thousands in numerals, breaking them up so they are easier to read. Other uses for commas are in dates (*March 19, 2016*), letter greetings (*Dear Sally,*), and in between cities and states (*Louisville, KY*).

Semicolons

A **semicolon** (;) is used to connect ideas in a sentence in some way. There are three main ways to use semicolons.

Link two independent clauses without the use of a coordinating conjunction:

> I was late for work again; I'm definitely going to get fired.

Link two independent clauses with a transitional word:

> The songs were all easy to play; therefore, he didn't need to spend too much time practicing.

Between items in a series that are already separated by commas or if necessary to separate lengthy items in a list:

> Starbucks has locations in Media, PA; Swarthmore, PA; and Morton, PA.
>
> Several classroom management issues presented in the study: the advent of a poor teacher persona in the context of voice, dress, and style; teacher follow-through from the beginning of the school year to the end; and the depth of administrative support, including ISS and OSS protocol.

Colons

A **colon** is used after an independent clause to present an explanation or draw attention to what comes next in the sentence. There are several uses.

Explanations of ideas:

> They soon learned the hardest part about having a new baby: sleep deprivation.

Lists of items:

> Shari picked up all the supplies she would need for the party: cups, plates, napkins, balloons, streamers, and party favors.

Time, subtitles, general salutations:

> The time is 7:15.
>
> I read a book entitled *Pluto: A Planet No More.*
>
> To whom it may concern:

Parentheses and Dashes

Parentheses are half-round brackets that look like this: ( ). They set off a word, phrase, or sentence that is an afterthought, explanation, or side note relevant to the surrounding text but not essential. A pair of commas is often used to set off this sort of information, but parentheses are generally used for information that would not fit well within a sentence or that the writer deems not important enough to be structurally part of the sentence.

> The picture of the heart (see above) shows the major parts you should memorize.
>
> Mount Everest is one of three mountains in the world that are over 28,000 feet high (K2 and Kanchenjunga are the other two).

See how the sentences above are complete without the parenthetical statements? In the first example, *see above* would not have fit well within the flow of the sentence. The second parenthetical statement could have been a separate sentence, but the writer deemed the information not pertinent to the topic.

The **em-dash** (—) is a mark longer than a hyphen used as a punctuation mark in sentences and to set apart a relevant thought. Even after plucking out the line separated by the dash marks, the sentence will be intact and make sense.

> Looking out the airplane window at the landmarks—Lake Clarke, Thompson Community College, and the bridge—she couldn't help but feel excited to be home.

The dashes use is similar to that of parentheses or a pair of commas. So, what's the difference? Many believe that using dashes makes the clause within them stand out while using parentheses is subtler. It's advised to not use dashes when commas could be used instead.

Ellipses

An **ellipsis** (…) consists of three handy little dots that can speak volumes on behalf of irrelevant material. Writers use them in place of words, lines, phrases, list content, or paragraphs that might just as easily have been omitted from a passage of writing. This can be done to save space or to focus only on the specifically relevant material.

> Exercise is good for some unexpected reasons. Watkins writes, "Exercise has many benefits such as…reducing cancer risk."

In the example above, the ellipsis takes the place of the other benefits of exercise that are more expected.

The ellipsis may also be used to show a pause in sentence flow.

> "I'm wondering...how this could happen," Dylan said in a soft voice.

Quotation Marks

Double **quotation marks** are used at the beginning and end of a direct quote. They are also used with certain titles and to indicate that a term being used is slang or referenced in the sentence. Quotation marks should not be used with an indirect quote. Single quotation marks are used to indicate a quote within a quote.

> Direct quote: "The weather is supposed to be beautiful this week," she said.
>
> Indirect quote: One of the customers asked if the sale prices were still in effect.
>
> Quote within a quote: "My little boy just said 'Mama, I want cookie,'" Maria shared.

Titles: Quotation marks should also be used to indicate titles of short works or sections of larger works, such as chapter titles. Other works that use quotation marks include poems, short stories, newspaper articles, magazine articles, web page titles, and songs.

> "The Road Not Taken" is my favorite poem by Robert Frost.
>
> "What a Wonderful World" is one of my favorite songs.

Specific or emphasized terms: Quotation marks can also be used to indicate a technical term or to set off a word that is being discussed in a sentence. Quotation marks can also indicate sarcasm.

> The new step, called "levigation," is a very difficult technique.
>
> He said he was "hungry" multiple times, but he only ate two bites.

Use with other punctuation: The use of quotation marks with other punctuation varies, depending on the role of the ending or separating punctuation.

In American English, periods and commas always go inside the quotation marks:

> "This is the last time you are allowed to leave early," his boss stated.
>
> The newscaster said, "We have some breaking news to report."

Question marks or exclamation points go inside the quotation marks when they are part of a direct quote:

> The doctor shouted, "Get the crash cart!"

When the question mark or exclamation point is part of the sentence, not the quote, it should be placed outside of the quotation marks:

> Was it Jackie that said, "Get some potatoes at the store"?

Apostrophes

This punctuation mark, the apostrophe (') is a versatile mark. It has several different functions:

- Quotes: Apostrophes are used when a second quote is needed within a quote.

  > In my letter to my friend, I wrote, "The girl had to get a new purse, and guess what Mary did? She said, 'I'd like to go with you to the store.' I knew Mary would buy it for her."

- Contractions: Another use for an apostrophe in the quote above is a contraction. *I'd is used for I would.*
- Possession: An apostrophe followed by the letter s shows possession (Mary's purse). If the possessive word is plural, the apostrophe generally just follows the word. Not all possessive pronouns require apostrophes.

  > The trees' leaves are all over the ground.

Hyphens

The **hyphen** (-) is a small hash mark that can be used to join words to show that they are linked.

Hyphenate two words that work together as a single adjective (a compound adjective).

> honey-covered biscuits

Some words always require hyphens, even if not serving as an adjective.

merry-go-round

Hyphens always go after certain prefixes like *anti-* & *all-*.

Hyphens should also be used when the absence of the hyphen would cause a strange vowel combination (*semi-engineer*) or confusion. For example, *re-collect* should be used to describe something being gathered twice rather than being written as *recollect*, which means to remember.

Subjects

Every sentence must include a subject and a verb. The **subject** of a sentence is who or what the sentence is about. It's often directly stated and can be determined by asking "Who?" or "What?" did the action:

Most sentences contain a direct subject, in which the subject is mentioned in the sentence.

Kelly mowed the lawn.

Who mowed the lawn? Kelly

The air-conditioner ran all night

What ran all night? the air-conditioner

The subject of imperative sentences is *you*, because imperative subjects are commands. The subject is implied because it is a command:

Go home after the meeting.

Who should go home after the meeting? you (implied)

In expletive sentences that start with "there are" or "there is," the subject is found after the predicate. The subject cannot be "there," so it must be another word in the sentence:

There is a cup sitting on the coffee table.

What is sitting on the coffee table? a cup

Simple and Complete Subjects

A **complete subject** includes the simple subject and all the words modifying it, including articles and adjectives. A **simple subject** is the single noun without its modifiers.

A warm, chocolate-chip cookie sat on the kitchen table.

Complete subject: *a warm, chocolate-chip cookie*

Simple subject: *cookie*

The words *a, warm, chocolate,* and *chip* all modify the simple subject *cookie*.

There might also be a **compound subject**, which would be two or more nouns without the modifiers.

A little girl and her mother walked into the shop.

Complete subject: *A little girl and her mother*

Compound subject: *girl, mother*

In this case, *the girl and her mother* are both completing the action of walking into the shop, so this is a compound subject.

Predicates

In addition to the subject, a sentence must also have a predicate. The **predicate** contains a verb and tells something about the subject. In addition to the verb, a predicate can also contain a direct or indirect object, object of a preposition, and other phrases.

The cats napped on the front porch.

In this sentence, cats is the subject because the sentence is about cats.

The **complete predicate** is everything else in the sentence: *napped on the front porch.* This phrase is the predicate because it tells us what the cats did.

This sentence can be broken down into a simple subject and predicate:

Cats napped.

In this sentence, *cats* is the simple subject, and *napped* is the *simple predicate*.

Although the sentence is very short and doesn't offer much information, it's still considered a complete sentence because it contains a subject and predicate.

Like a compound subject, a sentence can also have a **compound predicate**. This is when the subject is or does two or more things in the sentence.

This easy chair reclines and swivels.

In this sentence, *this easy chair* is the complete subject. *Reclines and swivels* shows two actions of the chair, so this is the compound predicate.

Subject-Verb Agreement

The subject of a sentence and its verb must agree. The cornerstone rule of subject-verb agreement is that subject and verb must agree in number. Whether the subject is singular or plural, the verb must follow suit.

Incorrect: The houses is new.

Correct: The houses are new.

Also Correct: The house is new.

In other words, a singular subject requires a singular verb; a plural subject requires a plural verb.

The words or phrases that come between the subject and verb do not alter this rule.

Incorrect: The houses built of brick is new.

Correct: The houses built of brick are new.

Incorrect: The houses with the sturdy porches is new.

Correct: The houses with the sturdy porches are new.

The subject will always follow the verb when a sentence begins with *here* or *there.* Identify these with care.

Incorrect: Here *is* the *houses* with sturdy porches.

Correct: Here *are* the *houses* with sturdy porches.

The subject in the sentences above is not *here*, it is *houses*. Remember, *here* and *there* are never subjects. Be careful that contractions such as *here's* or *there're* do not cause confusion!

Two subjects joined by *and* require a plural verb form, except when the two combine to make one thing:

Incorrect: Garrett and Jonathan is over there.

Correct: Garrett and Jonathan are over there.

Incorrect: Spaghetti and meatballs are a delicious meal!

Correct: Spaghetti and meatballs is a delicious meal!

In the example above, *spaghetti and meatballs* is a compound noun. However, *Garrett and Jonathan* is not a compound noun.

Two singular subjects joined by *or, either/or,* or *neither/nor* call for a singular verb form.

Incorrect: Butter or syrup are acceptable.

Correct: Butter or syrup is acceptable.

Plural subjects joined by *or*, *either/or*, or *neither/nor* are, indeed, plural.

The chairs *or* the boxes are being moved next.

If one subject is singular and the other is plural, the verb should agree with the closest noun.

Correct: The chair or the boxes are being moved next.

Correct: The chairs or the box is being moved next.

Some plurals of money, distance, and time call for a singular verb.

> Incorrect: Three dollars *are* enough to buy that.
>
> Correct: Three dollars *is* enough to buy that.

For words declaring degrees of quantity such as *many of, some of,* or *most of,* let the noun that follows of be the guide:

> Incorrect: Many of the books is in the shelf.
>
> Correct: Many of the books are in the shelf.
>
> Incorrect: Most of the pie *are* on the table.
>
> Correct: Most of the pie *is* on the table.

For indefinite pronouns like anybody or everybody, use singular verbs.

> Everybody *is* going to the store.

However, the pronouns *few, many, several, all, some,* and *both* have their own rules and use plural forms.

> Some *are* ready.

Some nouns like *crowd* and *congress* are called **collective nouns** and they require a singular verb form.

> Congress *is* in session.
>
> The news *is* over.

Books and movie titles, though, including plural nouns such as *Great Expectations*, also require a singular verb. Remember that only the subject affects the verb. While writing tricky subject-verb arrangements, say them aloud. Listen to them. Once the rules have been learned, one's ear will become sensitive to them, making it easier to pick out what's right and what's wrong.

Direct Objects

The **direct object** is the part of the sentence that receives the action of the verb. It is a noun and can usually be found after the verb. To find the direct object, first find the verb, and then ask the question *who* or *what* after it.

> The bear climbed the tree.
>
> What did the bear climb? *the tree*

Indirect Objects

An **indirect object** receives the direct object. It is usually found between the verb and the direct object. A strategy for identifying the indirect object is to find the verb and ask the questions *to whom/for whom* or *to what/ for what*.

Jane made her daughter a cake.

For whom did Jane make the cake? *her daughter*

Cake is the direct object because it is what Jane made, and *daughter* is the indirect object because she receives the cake.

Complements

A **complement** completes the meaning of an expression. A complement can be a pronoun, noun, or adjective. A verb complement refers to the direct object or indirect object in the sentence. An object complement gives more information about the direct object:

The magician got the kids excited.

Kids is the direct object, and *excited* is the object complement.

A **subject complement** comes after a linking verb. It is typically an adjective or noun that gives more information about the subject:

The king was noble and spared the thief's life.

Noble describes the *king* and follows the linking verb *was*.

Predicate Nouns

A **predicate noun** renames the subject:

John is a carpenter.

The subject is *John*, and the predicate noun is *carpenter*.

Predicate Adjectives

A **predicate adjective** describes the subject:

Margaret is beautiful.

The subject is *Margaret*, and the predicate adjective is *beautiful*.

Homonyms

Homonyms are words that sound the same but are spelled differently, and they have different meanings. There are several common homonyms that give writers trouble.

There, They're, and Their

The word *there* can be used as an adverb, adjective, or pronoun:

There are ten children on the swim team this summer.

I put my book over *there*, but now I can't find it.

The word *they're* is a contraction of the words *they* and *are*:

They're flying in from Texas on Tuesday.

The word *their* is a possessive pronoun:

I store *their* winter clothes in the attic.

Its and It's

Its is a possessive pronoun:

The cat licked *its* injured paw.

It's is the contraction for the words *it* and *is*:

It's unbelievable how many people opted not to vote in the last election.

Your and You're

Your is a possessive pronoun:

Can I borrow *your* lawnmower this weekend?

You're is a contraction for the words *you* and *are*:

You're about to embark on a fantastic journey.

To, Too, and Two

To is an adverb or a preposition used to show direction, relationship, or purpose:

We are going *to* New York.

They are going *to* see a show.

Too is an adverb that means more than enough, also, and very:

You have had *too* much candy.

We are on vacation that week, *too*.

Two is the written-out form of the numeral 2:

Two of the shirts didn't fit, so I will have to return them.

New and Knew

New is an adjective that means recent:

There's a *new* customer on the phone.

Knew is the past tense of the verb *know*:

I *knew* you'd have fun on this ride.

Affect and Effect

Affect and *effect* are complicated because they are used as both nouns and verbs, have similar meanings, and are pronounced the same.

	Affect	**Effect**
Noun Definition	emotional state	result
Noun Example	The patient's affect was flat.	The effects of smoking are well documented.
Verb Definition	to influence	to bring about
Verb Example	The pollen count affects my allergies.	The new candidate hopes to effect change.

Independent and Dependent Clauses

Independent and **dependent clauses** are strings of words that contain both a subject and a verb. An independent clause *can* stand alone as complete thought, but a dependent clause *cannot*. A dependent clause relies on other words to be a complete sentence.

Independent clause: The keys are on the counter.

Dependent clause: If the keys are on the counter

Notice that both clauses have a subject (*keys*) and a verb (*are*). The independent clause expresses a complete thought, but the word *if* at the beginning of the dependent clause makes it dependent on other words to be a complete thought.

Independent clause: If the keys are on the counter, please give them to me.

This example constitutes a complete sentence since it includes at least one verb and one subject and is a complete thought. In this case, the independent clause has two subjects (*keys* & an implied *you*) and two verbs (*are* & *give*).

Independent clause: I went to the store.

Dependent clause: Because we are out of milk,

Complete Sentence: Because we are out of milk, I went to the store.

Complete Sentence: I went to the store because we are out of milk.

Phrases

A **phrase** is a group of words that do not make a complete thought or a clause. They are parts of sentences or clauses. Phrases can be used as nouns, adjectives, or adverbs. A phrase does not contain both a subject and a verb.

Prepositional Phrases

A **prepositional phrase** shows the relationship between a word in the sentence and the object of the preposition. The object of the preposition is a noun that follows the preposition.

The orange pillows are on the couch.

On is the preposition, and *couch* is the object of the preposition.

She brought her friend with the nice car.

With is the preposition, and *car* is the object of the preposition. Here are some common prepositions:

about	as	at	after
by	for	from	in
of	on	to	with

Verbals and Verbal Phrases

Verbals are forms of verbs that act as other parts of speech. They can be used as nouns, adjectives, or adverbs. Though they are use verb forms, they are not to be used as the verb in the sentence. A word group that is based on a verbal is considered a **verbal phrase**. There are three major types of verbals: *participles, gerunds,* and *infinitives.*

Participles are verbals that act as adjectives. The present participle ends in *–ing*, and the past participle ends in *–d*, *-ed*, *-n*, or*-t*.

Verb	Present Participle	Past Participle
walk	walking	walked
share	sharing	shared

Participial phrases are made up of the participle and modifiers, complements, or objects.

Crying for most of an hour, the baby didn't seem to want to nap.

Having already taken this course, the student was bored during class.

Crying for most of an hour and *Having already taken this course* are the participial phrases.

Gerunds are verbals that are used as nouns and end in *–ing*. A gerund can be the subject or object of the sentence like a noun. Note that a present participle can also end in *–ing*, so it is important to distinguish between the two. The gerund is used as a noun, while the participle is used as an adjective.

Swimming is my favorite sport.

I wish I were sleeping.

A **gerund phrase** includes the gerund and any modifiers or complements, direct objects, indirect objects, or pronouns.

Cleaning the house is my least favorite weekend activity.

Cleaning the house is the gerund phrase acting as the subject of the sentence.

The most important goal this year is raising money for charity.

Raising money for charity is the gerund phrase acting as the direct object.

The police accused the woman of stealing the car.

The gerund phrase *stealing the car* is the object of the preposition in this sentence.

An **infinitive** is a verbal made up of the word *to* and a verb. Infinitives can be used as nouns, adjectives, or adverbs.

Examples: To eat, to jump, to swim, to lie, to call, to work

An **infinitive phrase** is made up of the infinitive plus any complements or modifiers. The infinitive phrase *to wait* is used as the subject in this sentence:

To wait was not what I had in mind.

The infinitive phrase *to sing* is used as the subject complement in this sentence:

Her dream is to sing.

The infinitive phrase *to grow* is used as an adverb in this sentence:

Children must eat to grow.

Appositive Phrases

An **appositive** is a noun or noun phrase that renames a noun that comes immediately before it in the sentence. An appositive can be a single word or several words. These phrases can be **essential** or **nonessential**. An essential appositive phrase is necessary to the meaning of the sentence and a nonessential appositive phrase is not. It is important to be able to distinguish these for purposes of comma use.

Essential: My sister Christina works at a school.

Naming which sister is essential to the meaning of the sentence, so no commas are needed.

Nonessential: My sister, who is a teacher, is coming over for dinner tonight.

Who is a teacher is not essential to the meaning of the sentence, so commas are required.

Absolute Phrases

An **absolute phrase** modifies a noun without using a conjunction. It is not the subject of the sentence and is not a complete thought on its own. Absolute phrases are set off from the independent clause with a comma.

Arms outstretched, she yelled at the sky.

All things considered, this has been a great day.

Four Types of Sentence Structures

A **simple sentence** has one independent clause.

I am going to win.

A **compound sentence** has two independent clauses. A conjunction—*for, and, nor, but, or, yet, so*—links them together. Note that each of the independent clauses has a subject and a verb.

I am going to win, but the odds are against me.

A **complex sentence** has one independent clause and one or more dependent clauses.

I am going to win, even though I don't deserve it.

Even though I don't deserve it is a dependent clause. It does not stand on its own. Some conjunctions that link an independent and a dependent clause are *although, because, before, after, that, when, which*, and *while*.

A **compound-complex sentence** has at least three clauses, two of which are independent and at least one that is a dependent clause.

While trying to dance, I tripped over my partner's feet, but I regained my balance quickly.

The dependent clause is *While trying to dance*.

Practice Test #1

AWA Prompt

The following paragraph appeared in the editorial section of a business magazine concerning on-call hours in retail stores:

> On-call hours in retail are when employees are scheduled to be "on-call" and are called into work at the last minute when the company needs them. All retail stores know that providing on-call hours helps the business save money by tailoring product sales to hours worked. Therefore, it is safe to say that implementing on-call hours is better for the business, which in turn is better for the workers.

In your essay response, evaluate the underlying logic of the given argument above. Analyze its use of evidence as well as its assumptions. You might also think about what kind of evidence might weaken or strengthen the argument. The essay should be approximately 400–600 words.

Integrated Reasoning

Multi-Source Reasoning

Prompt 1

Le Crème Ice Cream Company is popular in many Midwest states. The head of each department will be attending a meeting within the week. The following emails were circulated among the departments.

Email 1: Head of Marketing Department

The higher production costs of Blueberry Pie ice cream flavor have never been questioned due to it being one of the three most popular flavors Le Crème makes. With an increase in production costs, an adjustment in the marketing strategy could increase sales and profit. An idea for pairing the flavor with a charitable donation per purchase could increase the interest, provide a reason to relaunch the flavor in the media and perhaps expand its targeted consumer base.

Email 2: Head of Production Department

Le Crème's Blueberry Pie ice cream flavor has always been the most expensive to produce. This has been in part due to the cost of the blueberries and also from using the private company Crusts for the specially sized pieces of pie crust blended into the flavor that do not break apart. With the over flooding on the growing season in Michigan, the blueberries are expected to be scarce, and the contract with Crusts is up for renewal. Both factors make it likely that the cost of production will increase. It might be time to renegotiate with both suppliers or seek out new suppliers. In the interim, we may need to make it a "seasonal" flavor, halt production, or discontinue the flavor altogether if we cannot secure new contracts.

Email 3: Company Owner

My family carefully developed the Blueberry Pie ice cream flavor in the tradition that Le Crème had always followed. The flavor has been a great seller from its inception. Its sales do not vary based upon seasonal demand like some of the less popular flavors such as Eggnog or Pumpkin. The contracted companies were selected for the refined flavors and specific constitution of their products. Altering any part of the recipe will indeed change the entire flavor. This company was founded on tradition and quality. We are not willing to sacrifice either for a minor bump in the road.

1. It can be inferred that:

YES	NO	
O	O	All three individuals are focused on the continuance of the Blueberry Pie flavor.
O	O	The Blueberry Pie flavor was one of the company's original flavors.
O	O	The Blueberry Pie flavor is one of the company's top two selling flavors.

2. It can be inferred that:

YES	NO	
O	O	One of the primary points of discussion at the next meeting will be the future of the Blueberry Pie ice cream flavor.
O	O	The opinions of the owner and the head of marketing seem to be focused on finding a way to continue the Blueberry Pie ice cream flavor.

Prompt 2

A company has specific rules for paying employees premium pay (any time worked beyond the salaried 8-hour shift worked five days per week). The four types of premium pay and how they can be applied are described by the following excerpts.

Excerpt 1

Any employee who is scheduled to work any hours on a Sunday can claim Sunday differential. Sunday differential cannot be combined with any other type of premium pay, except night differential. Sunday differential is valued at 1.3 times an employee's base pay.

Excerpt 2

Night differential is offered in addition to regular base pay for time worked between the hours of 6:00 p.m. and 6:00 a.m. Night differential is valued at 1.25 times an employee's base pay.

Excerpt 3

Overtime is offered for hours worked beyond a regularly scheduled 8-hour shift that is worked five days per week. Overtime is valued at 1.5 times an employee's base pay.

Excerpt 4

Holiday pay is offered for hours, up to a maximum of 8 hours, that are worked on a federally designated holiday. Holiday pay cannot be combined with any other premium pay. Holiday pay is valued at two times an employee's base pay.

3. Based on the information provided, answer the following questions.

YES	NO	
O	O	An employee who works from 7:00 a.m. to 3:00 p.m. on July 4 (not falling on a Sunday) is eligible for 8 hours of premium pay.
O	O	An employee who works from 5:00 a.m. to 3:00 p.m. on July 4 (not falling on a Sunday) is eligible for 10 hours of holiday pay.

4. Employee A works the following shifts:

- Sunday: 9:00 a.m. to 5:00 p.m.
- Monday: 12:00 p.m. to 8:00 p.m.
- Tuesday: is scheduled off, but picks up a shift from 10:00 a.m. to 6:00 p.m.
- Wednesday: 6:00 a.m. to 2:00 p.m.
- Thursday: 5:00 a.m. to 1:00 p.m.
- Friday: 3:00 p.m. to 11:00 p.m.

If none of these days are a holiday, what is the maximum possible hours of premium pay the employee is eligible to receive?

a. 10 hours
b. 12 hours
c. 16 hours
d. 21 hours
e. 24 hours

Graphic Interpretation

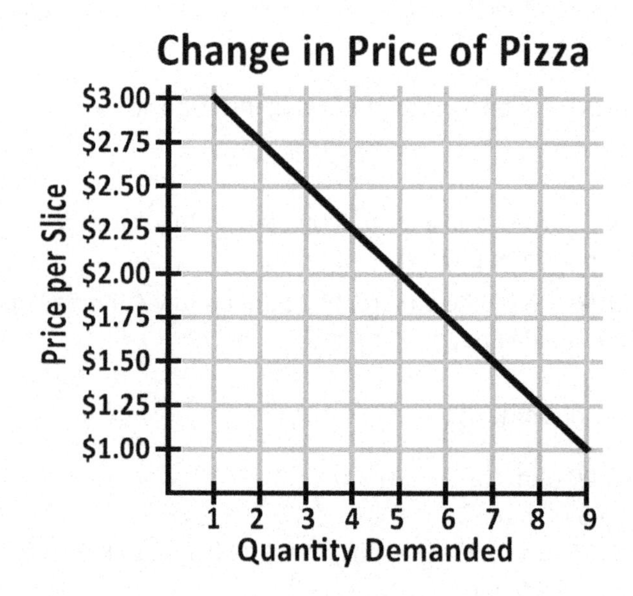

5. The average decrease in price for every increase in demand per slice is:
   a. $0.25
   b. $0.50
   c. $0.125
   d. $0.75

The price per slice and the quantity demanded have a:
   a. positive correlation
   b. negative correlation
   c. no apparent correlation

6. The diagram below shows the organization of rational and irrational numbers. From this, it can be determined that all are rational numbers, except:
   a. 0
   b. $\sqrt{3}$
   c. $\frac{1}{3}$
   d. $-\frac{1}{3}$

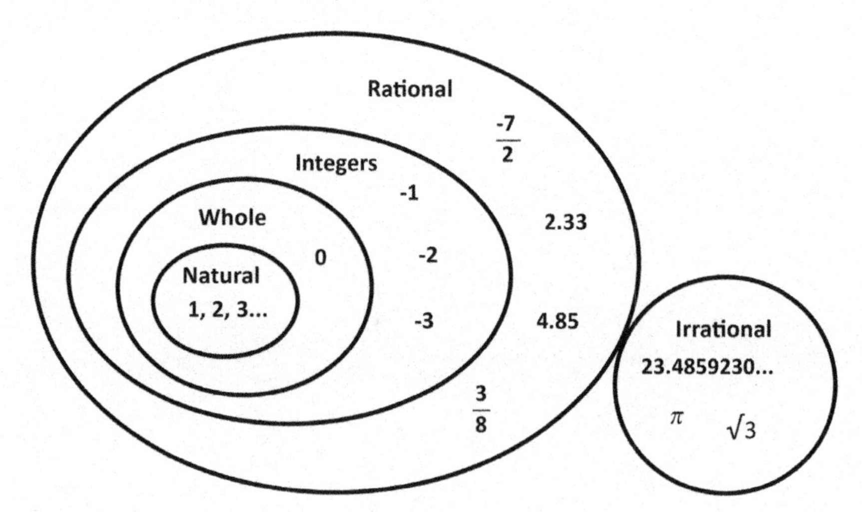

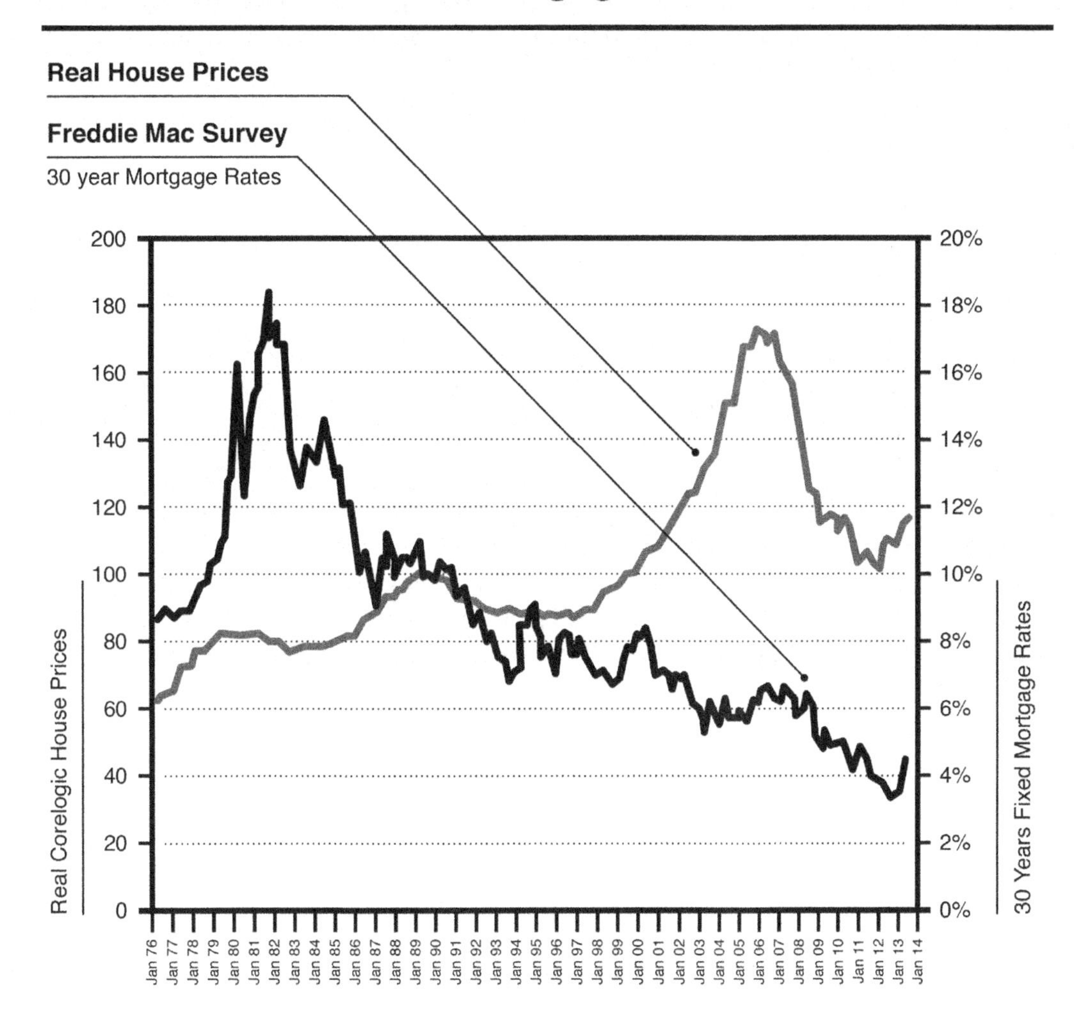

7. This graph shows the relation between real house prices and Freddie Mac Survey mortgage rates between January 1976 and January 2014.

This window of time maintains the largest consistent discrepancy between the real house prices and the survey house prices.

a. January 1981–January 1983
b. January 1980–January 1982
c. January 1998–January 2000
d. January 2006–January 2008

The real house price rises above the survey house price for the first time on approximately which date?

a. January 1995
b. January 1986
c. January 1989

Two-Part Analysis

8. Max can complete an analysis in 5 hours, while Dottie can compete the same analysis in 3 hours. Select a value for the time it would take them to finish the analysis if they worked together and by what difference this would be faster than Dottie's time.

Time Together	Difference	
O	O	1.625
O	O	1.875
O	O	1.125
O	O	0.625

9. Employees at Sales, Inc., must secure 10 sales per year. After their initial 10 sales, they receive an additional bonus of 5 percent of their base salary for every sale past the first 10 sales. For every sale over 20, an employee receives a bonus of $15,000 per sale. Employee A has a base salary of $54,000 and secures 12 sales. Employee B has a base salary of $62,000 and secures 21 sales. What equations could be used to calculate Employee B's pay and the difference between Employee A and Employee B's pay?

Employee B's Pay	Pay Difference Between Employee A and Employee B	
O	O	$54{,}000 + (a - 10)(0.05)(54{,}000)$
O	O	$62{,}000 + (b - 20)(15{,}000) + (b - 10)(0.05)(62{,}000)$
O	O	$54{,}000 + (a - 20)(15{,}000) + (a - 10)(.5)(54{,}000)$
O	O	$62{,}000 + (b - 20)(15{,}000) + (.05)(62{,}000)$
O	O	$62{,}000 + (b - 20)(15{,}000) + (b - 10)(0.05)(62{,}000) - (54{,}000 + (a - 10)(0.05)(54{,}000))$

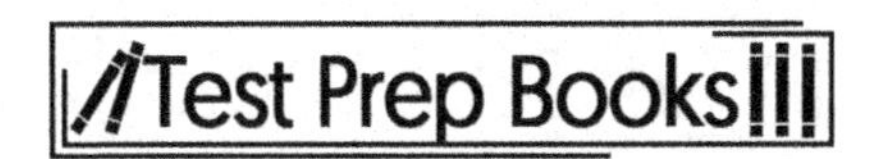

Table Analysis

Employee	Years	Position	Salary
Johnson	10	Warehouse	$44,000
Jones	15	HR	$45,000
Baker	19	Payroll	$42,000
Grande	9	Management	$81,000
Mays	4	Sales	$65,000
Innis	1	HR	$32,000
Richmond	6	Management	$92,000
Simpson	17	Marketing	$90,000
Douglas	22	HR	$50,000
Ryan	27	Warehouse	$49,000
Zimmers	2	Warehouse	$38,000

10.

TRUE	FALSE	
O	O	The average number of years in the warehouse division is 12 years.
O	O	The difference in average salary between the sales and management divisions is $21,500.
O	O	The division with the highest total number of years of experience in that position is the warehouse.

11.

TRUE FALSE

O O A plot of the years of experience versus salary would show a positive correlation.

Boys Game More Frequently

About how often do you play games?

	All teen gamers	Boy gamers	Girl gamers
Several times a day	13 %	19 %	6 %
About once a day	18	20	16
3 to 5 days a week	21	27	15
1 to 2 days a week	23	23	23
Every few weeks	15	7	23
Less often	10	4	17

12.

TRUE FALSE

O O The ratio of boy gamers to girl gamers who play about once a day is 5:4.

O O The percent of girl gamers compared to boy gamers in the categories of 3 to 5 days a week and 1 to 2 days a week is 72 percent.

O O The number of girl gamers tends to increase as the frequency of play during the week decreases.

Quantitative Reasoning

Problem Solving

1. Which of the following numbers has the greatest value?
   a. 1.4378
   b. 1.07548
   c. 1.43592
   d. 0.89409
   e. 1.42688

2. The value of 6×12 is the same as:
   a. $2 \times 4 \times 4 \times 2$
   b. $7 \times 4 \times 3$
   c. $6 \times 6 \times 3$
   d. $3 \times 2 \times 4 \times 3$
   e. $3 \times 4 \times 6 \times 2$

3. This chart indicates how many sales of CDs, vinyl records, and MP3 downloads occurred over the last year. Approximately what percentage of the total sales was from CDs?

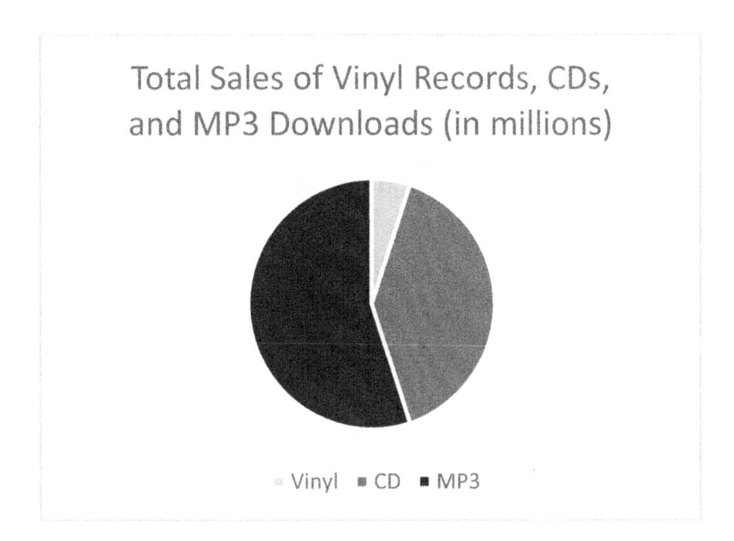

   a. 55%
   b. 25%
   c. 40%
   d. 5%
   e. 20%

4. After a 20% sale discount, Frank purchased a new refrigerator for $850. How much did he save from the original price?

 a. $170
 b. $212.50
 c. $105.75
 d. $200
 e. $150

5. A student gets an 85% on a test with 20 questions. How many answers did the student solve correctly?

 a. 16
 b. 15
 c. 18
 d. 19
 e. 17

6. Alan currently weighs 200 pounds, but he wants to lose weight to get down to 175 pounds. What is this difference in kilograms? (1 pound is approximately equal to 0.45 kilograms.)

 a. 9 kg
 b. 11.25 kg
 c. 78.75 kg
 d. 90 kg
 e. 25 kg

7. Johnny earns $2334.50 from his job each month. He pays $1437 for monthly expenses. Johnny is planning a vacation in 3 months that he estimates will cost $1750 total. How much will Johnny have left over from three months of saving once he pays for his vacation?

 a. $948.50
 b. $584.50
 c. $852.50
 d. $942.50
 e. $952.50

8. What is $\frac{420}{98}$ rounded to the nearest integer?

 a. 3
 b. 4
 c. 5
 d. 6
 e. 7

9. Dwayne has received the following scores on his math tests: 78, 92, 83, 97. What score must Dwayne get on his next math test to have an overall average of 90?

 a. 89
 b. 98
 c. 95
 d. 100
 e. 96

10. What is the overall median of Dwayne's current scores: 78, 92, 83, 97?
   a. 19
   b. 85
   c. 83
   d. 87.5
   e. 86

11. Solve the following:

$$(\sqrt{36} \times \sqrt{16}) - 3^2$$

   a. 30
   b. 21
   c. 15
   d. 13
   e. 16

12. In Jim's school, there are 3 girls for every 2 boys. There are 650 students in total. Using this information, how many students are girls?
   a. 260
   b. 130
   c. 65
   d. 390
   e. 225

13. Five of six numbers have a sum of 25. The average of all six numbers is 6. What is the sixth number?
   a. 8
   b. 12
   c. 13
   d. 10
   e. 11

14. Kimberley earns $10 an hour babysitting, and after 10 p.m., she earns $12 an hour, with the amount paid being rounded to the nearest hour accordingly. On her last job, she worked from 5:30 p.m. to 11 p.m. In total, how much did Kimberley earn on her last job?
   a. $45
   b. $57
   c. $62
   d. $42
   e. $67

15. Arrange the following numbers from least to greatest value:
$0.85, \frac{4}{5}, \frac{2}{3}, \frac{91}{100}$

a. $0.85, \frac{4}{5}, \frac{2}{3}, \frac{91}{100}$

b. $\frac{4}{5}, 0.85, \frac{91}{100}, \frac{2}{3}$

c. $\frac{2}{3}, \frac{4}{5}, 0.85, \frac{91}{100}$

d. $0.85, \frac{91}{100}, \frac{4}{5}, \frac{2}{3}$

e. $\frac{4}{5}, \frac{2}{3}, 0.85, \frac{91}{100}$

16. Keith's bakery had 252 customers go through its doors last week. This week, that number increased to 378. Express this increase as a percentage.
   a. 26%
   b. 50%
   c. 35%
   d. 12%
   e. 28%

17. The following graph compares the various test scores of the top three students in each of these teacher's classes. Based on the graph, which teacher's students had the lowest range of test scores?

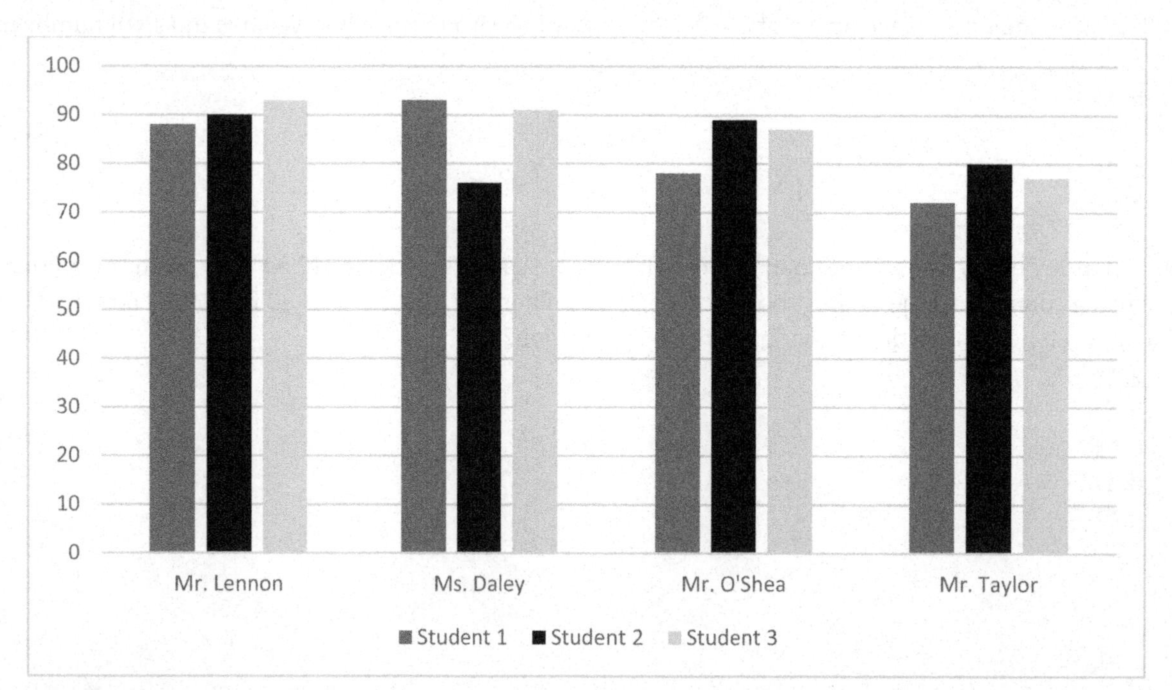

   a. Mr. Lennon
   b. Mr. O'Shea
   c. Mr. Taylor
   d. Ms. Daley
   e. Ms. Carter

18. Four people split a bill. The first person pays for $\frac{1}{5}$, the second person pays for $\frac{1}{4}$, and the third person pays for $\frac{1}{3}$. What fraction of the bill does the fourth person pay?

a. $\frac{13}{60}$

b. $\frac{47}{60}$

c. $\frac{1}{4}$

d. $\frac{4}{15}$

e. $\frac{1}{2}$

19. Simplify the following expression:

$$4\frac{2}{3} - 3\frac{4}{9}$$

a. $1\frac{1}{3}$

b. $1\frac{2}{9}$

c. 1

d. $1\frac{2}{3}$

e. $1\frac{4}{9}$

20. A closet is filled with red, blue, and green shirts. If $\frac{1}{3}$ of the shirts are green and $\frac{2}{5}$ are red, what fraction of the shirts are blue?

a. $\frac{4}{15}$

b. $\frac{1}{5}$

c. $\frac{7}{15}$

d. $\frac{1}{2}$

e. $\frac{2}{3}$

Data Sufficiency

Decide whether the data given in the statements are sufficient to answer the question.

21. What is the value of angle B in right triangle ABC?

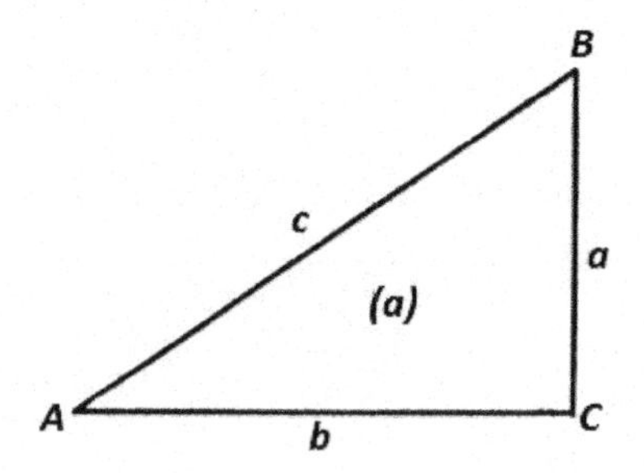

1) $c = 5$
2) Angle A measures 30 degrees.

a. Statement (1) ALONE is sufficient, but statement (2) alone is not sufficient to answer the question asked.
b. Statement (2) ALONE is sufficient, but statement (1) alone is not sufficient to answer the question asked.
c. BOTH statements (1) and (2) TOGETHER are sufficient to answer the question asked, but NEITHER statement ALONE is sufficient to answer the question asked.
d. EACH statement ALONE is sufficient to answer the question asked.
e. Statements (1) and (2) TOGETHER are NOT sufficient to answer the question asked, and additional data specific to the problem are needed.

22. What is the measure of angle 2?

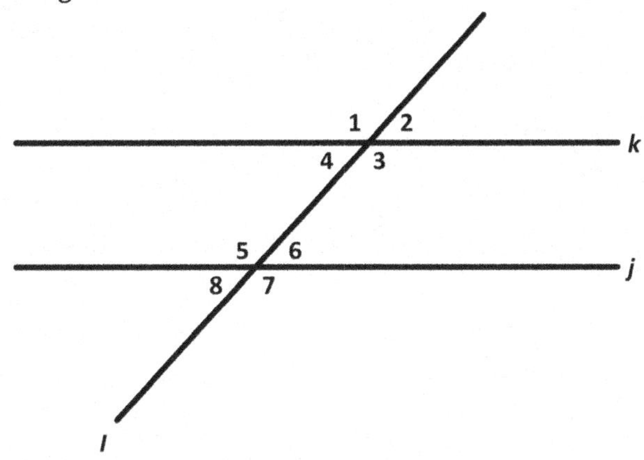

1) *k* and *j* are parallel lines, with transversal line *l*.
2) Angle 8 measures 53 degrees.

a. Statement (1) ALONE is sufficient, but statement (2) alone is not sufficient to answer the question asked.
b. Statement (2) ALONE is sufficient, but statement (1) alone is not sufficient to answer the question asked.
c. BOTH statements (1) and (2) TOGETHER are sufficient to answer the question asked, but NEITHER statement ALONE is sufficient to answer the question asked.
d. EACH statement ALONE is sufficient to answer the question asked.
e. Statements (1) and (2) TOGETHER are NOT sufficient to answer the question asked, and additional data specific to the problem are needed.

23. What is the numerical value of $\left(x - \frac{1}{2}\right)^2$?

1) $x^2 + x - 2 = 0$
2) $x > 0$

a. Statement (1) ALONE is sufficient, but statement (2) alone is not sufficient to answer the question asked.
b. Statement (2) ALONE is sufficient, but statement (1) alone is not sufficient to answer the question asked.
c. BOTH statements (1) and (2) TOGETHER are sufficient to answer the question asked, but NEITHER statement ALONE is sufficient to answer the question asked.
d. EACH statement ALONE is sufficient to answer the question asked.
e. Statements (1) and (2) TOGETHER are NOT sufficient to answer the question asked, and additional data specific to the problem are needed.

24. A motorcycle race consists of three legs. On the first two legs of the race, a motorcycle travels an average of 50 miles per hour. On the last leg of the race, it travels an average of 75 miles per hour. What is this motorcycle's average speed for the entire race in miles per hour?

1) All three legs are of equal length.
2) The total distance is 450 miles.

a. Statement (1) ALONE is sufficient, but statement (2) alone is not sufficient to answer the question asked.
b. Statement (2) ALONE is sufficient, but statement (1) alone is not sufficient to answer the question asked.
c. BOTH statements (1) and (2) TOGETHER are sufficient to answer the question asked, but NEITHER statement ALONE is sufficient to answer the question asked.
d. EACH statement ALONE is sufficient to answer the question asked.
e. Statements (1) and (2) TOGETHER are NOT sufficient to answer the question asked, and additional data specific to the problem are needed.

25. What is 40 percent of 20w?

1) $w \neq 0$
2) w is positive

a. Statement (1) ALONE is sufficient, but statement (2) alone is not sufficient to answer the question asked.
b. Statement (2) ALONE is sufficient, but statement (1) alone is not sufficient to answer the question asked.
c. BOTH statements (1) and (2) TOGETHER are sufficient to answer the question asked, but NEITHER statement ALONE is sufficient to answer the question asked.
d. EACH statement ALONE is sufficient to answer the question asked.
e. Statements (1) and (2) TOGETHER are NOT sufficient to answer the question asked, and additional data specific to the problem are needed.

26. If $a^2 - b^2 = 12$, what is the value of ab?

1) $(a - b) = 4$
2) $a \neq 0$

a. Statement (1) ALONE is sufficient, but statement (2) alone is not sufficient to answer the question asked.
b. Statement (2) ALONE is sufficient, but statement (1) alone is not sufficient to answer the question asked.
c. BOTH statements (1) and (2) TOGETHER are sufficient to answer the question asked, but NEITHER statement ALONE is sufficient to answer the question asked.
d. EACH statement ALONE is sufficient to answer the question asked.
e. Statements (1) and (2) TOGETHER are NOT sufficient to answer the question asked, and additional data specific to the problem are needed.

27. How many cans of paint can Jessica buy with $16?

1) One color was $1 more per can than the other.
2) She only bought red cans and blue cans, and she bought two more blue cans than red cans.

a. Statement (1) ALONE is sufficient, but statement (2) alone is not sufficient to answer the question asked.
b. Statement (2) ALONE is sufficient, but statement (1) alone is not sufficient to answer the question asked.
c. BOTH statements (1) and (2) TOGETHER are sufficient to answer the question asked, but NEITHER statement ALONE is sufficient to answer the question asked.
d. EACH statement ALONE is sufficient to answer the question asked.
e. Statements (1) and (2) TOGETHER are NOT sufficient to answer the question asked, and additional data specific to the problem are needed.

28. Choose the correct statement(s) for the following equation: $\sqrt{x^2 - 2xy + y^2} = 16$

1) $x > 0$
2) $x = 5y$

a. Statement (1) ALONE is sufficient, but statement (2) alone is not sufficient to answer the question asked.
b. Statement (2) ALONE is sufficient, but statement (1) alone is not sufficient to answer the question asked.
c. BOTH statements (1) and (2) TOGETHER are sufficient to answer the question asked, but NEITHER statement ALONE is sufficient to answer the question asked.
d. EACH statement ALONE is sufficient to answer the question asked.
e. Statements (1) and (2) TOGETHER are NOT sufficient to answer the question asked, and additional data specific to the problem are needed.

29. $a + b = 9$

$a - c = 14$

Find $c - b$

1) $a \neq 0$
2) $a = 10$

a. Statement (1) ALONE is sufficient, but statement (2) alone is not sufficient to answer the question asked.
b. Statement (2) ALONE is sufficient, but statement (1) alone is not sufficient to answer the question asked.
c. BOTH statements (1) and (2) TOGETHER are sufficient to answer the question asked, but NEITHER statement ALONE is sufficient to answer the question asked.
d. EACH statement ALONE is sufficient to answer the question asked.
e. Statements (1) and (2) TOGETHER are NOT sufficient to answer the question asked, and additional data specific to the problem are needed.

30. An employee receives 12 percent commission on every sale. How much was the sale price of sale x?

1) Commission = $156
2) Sale x + Sale y = $20,000

a. Statement (1) ALONE is sufficient, but statement (2) alone is not sufficient to answer the question asked.
b. Statement (2) ALONE is sufficient, but statement (1) alone is not sufficient to answer the question asked.
c. BOTH statements (1) and (2) TOGETHER are sufficient to answer the question asked, but NEITHER statement ALONE is sufficient to answer the question asked.
d. EACH statement ALONE is sufficient to answer the question asked.
e. Statements (1) and (2) TOGETHER are NOT sufficient to answer the question asked, and additional data specific to the problem are needed.

31. Points L and M lie on line KN. The length of line KN is 30 units long, and KM is 16 units long. How many units long is LM?

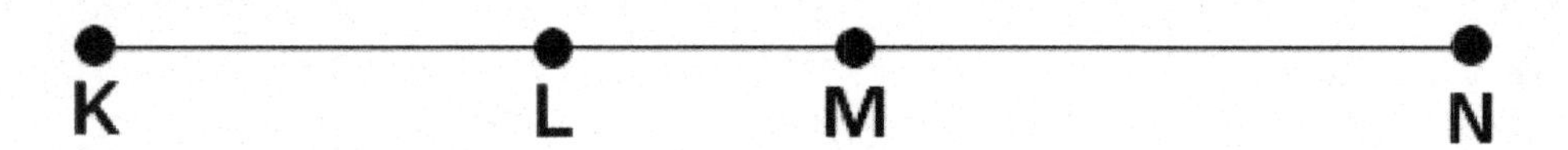

1) Points K and N lie on line KN.
2) The length of LN is 20 units long.

a. Statement (1) ALONE is sufficient, but statement (2) alone is not sufficient to answer the question asked.
b. Statement (2) ALONE is sufficient, but statement (1) alone is not sufficient to answer the question asked.
c. BOTH statements (1) and (2) TOGETHER are sufficient to answer the question asked, but NEITHER statement ALONE is sufficient to answer the question asked.
d. EACH statement ALONE is sufficient to answer the question asked.
e. Statements (1) and (2) TOGETHER are NOT sufficient to answer the question asked, and additional data specific to the problem are needed.

32. A 50-foot ladder is leaned up against the side of a building at an angle. The side of the building is perpendicular to the ground. How far up the building does the ladder reach?

1) The ladder makes a 53.1 degree angle with the ground.
2) The ladder makes a 36.9 degree angle with the building.

a. Statement (1) ALONE is sufficient, but statement (2) alone is not sufficient to answer the question asked.
b. Statement (2) ALONE is sufficient, but statement (1) alone is not sufficient to answer the question asked.
c. BOTH statements (1) and (2) TOGETHER are sufficient to answer the question asked, but NEITHER statement ALONE is sufficient to answer the question asked.
d. EACH statement ALONE is sufficient to answer the question asked.
e. Statements (1) and (2) TOGETHER are NOT sufficient to answer the question asked, and additional data specific to the problem are needed.

33. Of the 804 graduating seniors, a portion are going on to salaried positions after college. Approximately $\frac{1}{4}$ of those positions will be in state. What is the closest estimate for how many graduates taking salaried positions will be staying in state?

1) $\frac{2}{5}$ of the 804 are going on to salaried positions.
2) $\frac{3}{4}$ of the salaried positions are out of state.
a. Statement (1) ALONE is sufficient, but statement (2) alone is not sufficient to answer the question asked.
b. Statement (2) ALONE is sufficient, but statement (1) alone is not sufficient to answer the question asked.
c. BOTH statements (1) and (2) TOGETHER are sufficient to answer the question asked, but NEITHER statement ALONE is sufficient to answer the question asked.
d. EACH statement ALONE is sufficient to answer the question asked.
e. Statements (1) and (2) TOGETHER are NOT sufficient to answer the question asked, and additional data specific to the problem are needed.

34. A rectangle's length is twice its width. What is the value of the area of the rectangle?

1) The width of the rectangle is 2 units.
2) The perimeter of the rectangle is 12 units.

a. Statement (1) ALONE is sufficient, but statement (2) alone is not sufficient to answer the question asked.
b. Statement (2) ALONE is sufficient, but statement (1) alone is not sufficient to answer the question asked.
c. BOTH statements (1) and (2) TOGETHER are sufficient to answer the question asked, but NEITHER statement ALONE is sufficient to answer the question asked.
d. EACH statement ALONE is sufficient to answer the question asked.
e. Statements (1) and (2) TOGETHER are NOT sufficient to answer the question asked, and additional data specific to the problem are needed.

35. A bag contains red, blue, and white marbles. What are the chances a red marble is drawn out of a bag on the first try?

1) The bag contains a total of 12 marbles.
2) There is a $\frac{1}{2}$ chance of drawing a white marble and a $\frac{1}{4}$ chance of drawing a blue marble.

a. Statement (1) ALONE is sufficient, but statement (2) alone is not sufficient to answer the question asked.
b. Statement (2) ALONE is sufficient, but statement (1) alone is not sufficient to answer the question asked.
c. BOTH statements (1) and (2) TOGETHER are sufficient to answer the question asked, but NEITHER statement ALONE is sufficient to answer the question asked.
d. EACH statement ALONE is sufficient to answer the question asked.
e. Statements (1) and (2) TOGETHER are NOT sufficient to answer the question asked, and additional data specific to the problem are needed.

36. A local deli offers customers a choice of meat, bread, and cheese for their sandwiches. How many different combinations are there for a sandwich selection?

1) Wheat bread is the most popular choice of bread.
2) There are three types of meat, three types of cheese, and five types of bread.

a. Statement (1) ALONE is sufficient, but statement (2) alone is not sufficient to answer the question asked.
b. Statement (2) ALONE is sufficient, but statement (1) alone is not sufficient to answer the question asked.
c. BOTH statements (1) and (2) TOGETHER are sufficient to answer the question asked, but NEITHER statement ALONE is sufficient to answer the question asked.
d. EACH statement ALONE is sufficient to answer the question asked.
e. Statements (1) and (2) TOGETHER are NOT sufficient to answer the question asked, and additional data specific to the problem are needed.

37. A farmer owns two (non-adjacent) plots of land, which he wishes to fence. How much fencing does he need, in feet, in total to enclose each of the two plots?

1) The area of one is 1000 square feet.
2) The area of the other is 10 square feet.

a. Statement (1) ALONE is sufficient, but statement (2) alone is not sufficient to answer the question asked.
b. Statement (2) ALONE is sufficient, but statement (1) alone is not sufficient to answer the question asked.
c. BOTH statements (1) and (2) TOGETHER are sufficient to answer the question asked, but NEITHER statement ALONE is sufficient to answer the question asked.
d. EACH statement ALONE is sufficient to answer the question asked.
e. Statements (1) and (2) TOGETHER are NOT sufficient to answer the question asked, and additional data specific to the problem are needed.

Verbal Reasoning

Reading Comprehension

Questions 1–4 are based on the following passage:

> The town of Alexandria, Virginia was founded in 1749. Between the years 1810 and 1861, this thriving seaport was the ideal location for slave owners such as Joseph Bruin, Henry Hill, Isaac Franklin, and John Armfield to build several slave trade office structures, including slave holding areas. After 1830, when the manufacturing-based economy slowed down in Virginia, slaves were traded to plantations in the Deep South, in Alabama, Mississippi, and Louisiana. Joseph Bruin, one of the most notorious of the slave traders operating in Alexandria, alone purchased hundreds of slaves from 1844 to 1861. Harriet Beecher Stowe claimed that the horrible slave traders mentioned in her novel, *Uncle Tom's Cabin*, are reminiscent of the coldhearted Joseph Bruin. The Franklin and Armfield Office was known as one of the largest slave trading companies in the country up to the end of the Civil War period. Slaves, waiting to be traded, were held in a two-story slave pen built behind the Franklin and Armfield Office structure on Duke Street in Alexandria. Yet, many people fought to thwart these traders and did everything they could to rescue and free slaves. Two Christian African American slave sisters, with the help of northern abolitionists who bought their freedom, escaped Bruin's plan to sell them into southern prostitution. In 1861, Joseph Bruin was captured and imprisoned and his property confiscated. The Bruin Slave Jail became the Fairfax County courthouse until 1865. The original Franklin and Armfield Office building still stands in Virginia and is registered in the National Register of Historic Places. The Bruin Slave Jail is still standing on Duke Street in Alexandria, but is not open to the public. The history of the slave trading enterprise is preserved and presented to the public by the Northern Virginia Urban League.

1. Based on the above passage, which of the following statements about the town of Alexandria are true?
   a. Alexandria was a seaport town, which could not prosper, even with the advent of a slave trade business, because the manufacturing industry was not enough to stabilize the economy.
   b. Slave traders such as Joseph Bruin, Henry Hill, Isaac Franklin, and John Armfield rented both slave trade office buildings and slave holding buildings from landlords of Old Town, Alexandria.
   c. For over fifteen years, Joseph Bruin, a notorious slave trader, probably the one characterized in *Uncle Tom's Cabin*, bought hundreds of slaves with the intention of sending the purchased slaves to southern states such as Alabama, Mississippi, and Louisiana.
   d. The Bruin Slave Jail is open to the public; the building is located in downtown Alexandria, and still stands in Virginia. The jail is registered in the National Register of Historic Places. The history of the slave trading enterprise is preserved and presented to the public by the Northern Virginia Urban League.
   e. Isaac Franklin and John Armfield's slave-trade office structures, including slave holding areas in downtown Alexandria, did not remain open for their slave trade business until the end of the Civil War.

2. The passage about the Alexandria slave trade business suggests that which of the following statements can be regarded as true?
   a. The lucrative seaport town of Alexandria was supported by successful slave trade businesses of men like Joseph Bruin, Henry Hill, Isaac Franklin, and John Armfield, who bought slaves and sold them to the plantations in the Deep South.
   b. Joseph Bruin, a highly respected Alexandrian businessman, ran a slave trade business in downtown Alexandria until the business closed its doors at the end of the Civil War.
   c. The Franklin and Armfield Office was built by Isaac Franklin and John Armfield. Slaves, waiting to be traded, were held in a four-story slave pen built behind the Franklin and Armfield Office structure on Duke Street in Alexandria.
   d. When the Confederate Army positioned its command in Alexandria and closed slave traders' businesses, the Franklin and Armfield slave pen became the Fairfax County courthouse and was used to hold Union soldiers.
   e. The literature of the slave trading enterprise, like *Uncle Tom's Cabin*, is being preserved and presented to the public by the Northern Virginia Urban League.

3. Which of the following statements can be inferred to be accurate, based on the information provided in the passage?
   a. The town of Alexandria, founded in 1810, became one of the most infamous slave trading markets in the country.
   b. Harriet Beecher Stowe was an escaped slave who was held in the Franklin and Armfield slave pen, located on Duke Street in Alexandria. To avoid a life as a prostitute, Miss Stowe tried to escape from the control of Joseph Bruin, whose surly characteristics surfaced in her classic book, *Uncle Tom's Cabin*.
   c. Northern abolitionists were not known to help runaway slaves escape the hands of their notorious owners.
   d. The Bruin Slave Jail, located in downtown Alexandria, still stands today, although it is not open for public viewing.
   e. For convenience, the slave traders took their slaves to nearby Annapolis, Maryland, because the cost of shipping them from there was less than the cost of shipping them from Alexandria.

4. Which of the following statements best illustrates the author's intended main point or thesis?
   a. Two Christian African American slave sisters, with the help of northern abolitionists who bought their freedom, escaped Bruin's plan to sell them into southern prostitution.
   b. The town of Alexandria, a thriving seaport founded in 1749, was the location for several lucrative slave trading companies from 1810 to 1861.
   c. After the start of the Civil War, Joseph Bruin was captured and his jail was no longer used for his slave trade business.
   d. The Bruin Slave Jail is still standing on Duke Street in Alexandria, but is not open to the public.
   e. In 1861, the Bruin Slave Jail in Alexandria became the Fairfax County courthouse.

Questions 5–9 are based on the following passage:

> Becoming a successful leader in today's industry, government, and nonprofit sectors requires more than a high intelligence quotient (IQ). Emotional Intelligence (EI) includes developing the ability to know one's own emotions, to regulate impulses and emotions, and to use interpersonal communication skills with ease while dealing with other people. A combination of knowledge, skills, abilities, and mature emotional intelligence (EI) reflects the most effective leadership recipe. Successful leaders sharpen more than their talents and IQ levels; they practice

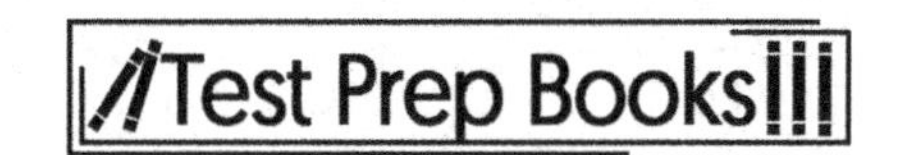

the basic features of emotional intelligence. Some of the hallmark traits of a competent, emotionally intelligent leader include self-efficacy, drive, determination, collaboration, vision, humility, and openness to change. An unsuccessful leader exhibits opposite leadership traits: unclear directives, inconsistent vision and planning strategies, disrespect for followers, incompetence, and an uncompromising transactional leadership style. There are ways to develop emotional intelligence for the person who wants to improve their leadership style. For example, an emotionally intelligent leader creates an affirmative environment by incorporating collaborative activities, using professional development training for employee self-awareness, communicating clearly about the organization's vision, and developing a variety of resources for working with emotions. Building relationships outside the institution with leadership coaches and with professional development trainers can also help leaders who want to grow their leadership success. Leaders in today's work environment need to strive for a combination of skill, knowledge, and mature emotional intelligence to lead followers to success and to promote the vision and mission of their respective institutions.

5. The passage suggests that the term *emotional intelligence (EI)* can be defined as which of the following?

a. A combination of knowledge, skills, abilities, and mature emotional intelligence reflects the most effective EI leadership recipe.
b. An emotionally intelligent leader creates an affirmative environment by incorporating collaborative activities, using professional development training for employee self-awareness, communicating clearly about the organization's vision, and developing a variety of resources for working with emotions.
c. EI includes developing the ability to know one's own emotions, to regulate impulses and emotions, and to use interpersonal communication skills with ease while dealing with other people.
d. Becoming a successful leader in today's industry, government, and nonprofit sectors requires more than a high IQ.
e. An EI leader exhibits the following leadership traits: unclear directives, inconsistent vision and planning strategies, disrespect for followers, incompetence, and uncompromising transactional leadership style.

6. Based on the information in the passage, a successful leader must have a high EI quotient.

a. The above statement can be supported by the fact that Daniel Goldman conducted a scientific study.
b. The above statement can be supported by the example that emotionally intelligent people are highly successful leaders.
c. The above statement is not supported by the passage.
d. The above statement is supported by the illustration that claims, "Leaders in today's work environment need to strive for a combination of skill, knowledge, and mature emotional intelligence to lead followers to success and to promote the vision and mission of their respective institutions."
e. The above statement can be inferred because emotionally intelligent people obviously make successful leaders.

7. According to the passage above, some of the characteristics of an unsuccessful leader include which of the following?
   a. Talent, IQ level, and abilities
   b. Humility, knowledge, and skills
   c. Loud, demeaning actions toward female employees
   d. Outdated technological resources and strategies
   e. Transactional leadership style

8. According to the passage above, which of the following must be true?
   a. The leader exhibits a healthy work/life balance lifestyle.
   b. The leader is uncompromising in transactional directives for all employees, regardless of status.
   c. The leader learns to strategize using future trends analysis to create a five-year plan.
   d. The leader uses a combination of skill, knowledge, and mature reasoning to make decisions.
   e. The leader continually tries to improve their EI test quotient by studying the intelligence quotient of other successful leaders.

9. According to the passage above, which of the following choices are true?
   a. To be successful, leaders in the nonprofit sector need to develop neurological intelligence.
   b. It is not necessary for military leaders to develop emotional intelligence because they prefer a transactional leadership style.
   c. Leadership coaches cannot add value to someone who is developing their emotional intelligence.
   d. Humility is a valued character value; however, it is not necessarily a trademark of an emotionally intelligent leader.
   e. If a leader does not have the level of emotional intelligence required for a certain job, they are capable of increasing emotional intelligence.

Questions 10–14 are based on the following passage:

> Learning how to write a ten-minute play may seem like a monumental task at first; but, if you follow a simple creative writing strategy, similar to writing a narrative story, you will be able to write a successful drama. The first step is to open your story as if it is a puzzle to be solved. This will allow the reader a moment to engage with the story and to mentally solve the story with you, the author. Immediately provide descriptive details that steer the main idea, the tone, and the mood according to the overarching theme you have in mind. For example, if the play is about something ominous, you may open Scene One with a thunderclap. Next, use dialogue to reveal the attitudes and personalities of each of the characters who have a key part in the unfolding story. Keep the characters off balance in some way to create interest and dramatic effect. Maybe what the characters say does not match what they do. Show images on stage to speed up the narrative; remember, one picture speaks a thousand words. As the play progresses, the protagonist must cross the point of no return in some way; this is the climax of the story. Then, as in a written story, you create a resolution to the life-changing event of the protagonist. Let the characters experience some kind of self-discovery that can be understood and appreciated by the patient audience. Finally, make sure all things come together in the end so that every detail in the play makes sense right before the curtain falls.

10. Based on the passage above, which of the following statements is FALSE?
    a. Writing a ten-minute play may seem like an insurmountable task.
    b. Providing descriptive details is not necessary until after the climax of the story line.
    c. Engaging the audience by jumping into the story line immediately helps them solve the story's developing ideas with you, the writer.
    d. Descriptive details give clues to the play's intended mood and tone.
    e. The introduction of a ten-minute play does not need to open with a lot of coffee pouring or cigarette smoking to introduce the scenes. The action can get started right away.

11. Based on the passage above, which of the following is true?
    a. The class of eighth graders quickly learned that it is not that difficult to write a ten-minute play.
    b. The playwrights of the twenty-first century all use the narrative writing basic feature guide to outline their initial scripts.
    c. In order to follow a simple structure, a person can write a ten-minute play based on some narrative writing features.
    d. Women find playwriting easier than men because they are used to communicating in writing.
    e. The structure of writing a poem is similar to that of play writing and of narrative writing.

12. Based on your understanding of the passage above, it can be assumed that which of the following statements are true?
    a. One way to reveal the identities and nuances of the characters in a play is to use props.
    b. Characters should follow predictable routes in the challenge presented in the unfolding narrative, so the audience may easily follow the sequence of events.
    c. Using images in the stage design is an important element of creating atmosphere and meaning for the drama.
    d. There is no need for the protagonist to come to terms with a self-discovery; he or she simply needs to follow the prescription for life lived as usual.
    e. It is perfectly fine to avoid serious consequences for the actors of a ten-minute play because there is not enough time to unravel perils.

13. In the passage above, the writer suggests that writing a ten-minute play is accessible for a novice playwright because of which of the following reasons?
    a. It took the author of the passage only one week to write his first play.
    b. The format follows similar strategies of writing a narrative story.
    c. There are no particular themes or points to unravel; a playwright can use a stream of consciousness style to write a play.
    d. Dialogue that reveals the characters' particularities is uncommonly simple to write.
    e. The characters of a ten-minute play wrap up the action simply by revealing their ideas in a monologue.

14. Based on the passage above, which basic feature of narrative writing is NOT mentioned with respect to writing a ten-minute play?
    a. Character development
    b. Descriptive details
    c. Dialogue
    d. Mood and tone
    e. Style

Critical Reasoning

15. Dalton: Agribusiness is an industry that's polluting the world and causing greenhouse emissions as quickly as the oil and automobile industries combined. In order to reduce global warming, we should think about eating less of the types of foods that agribusiness sells.

Alexis: If we want to reduce global warming, we should seriously think about what kind of cars we drive.

Alexis' response to Dalton serves to do which of the following?

a. Serves to reinforce Dalton's claims of the causes of global warming.
b. Strengthens Dalton's argument by adding to it.
c. Questions whether Dalton knows what he's talking about.
d. Backs up Dalton's claims with more evidence of the causes of global warming.
e. Suggest an alternative factor to the problem of global warming.

16. Shelby has fifteen years of coding experience. I thought that she would be the perfect fit for your company, since you are looking for a proficient coder.

The speaker assumes which of the following?

a. That Shelby is an excellent coder.
b. That the company is looking for someone with fifteen years of experience.
c. That Shelby will be a bad fit for the company.
d. That fifteen years of experience makes someone a proficient coder.
e. That Shelby is the best candidate for the job.

17. The "complete protein" is a source of protein that contains all nine of the essential amino acids necessary for our dietary needs. The complete protein is found in meat. Therefore, everyone should eat meat to fulfill their protein needs.

Which of the following, if true, would most weaken the above argument?

a. Grass-fed beef is the most ethical way to buy meat to fulfill your protein needs.
b. Legumes such as beans, seeds, or lentils mixed with grains such as pasta, rice, or corn creates a complete protein.
c. Amino acids are necessary for humans to ingest; the body cannot make its own amino acids.
d. Fruits and vegetables contain essential fiber, vitamins, and minerals that help our bodies fight disease.
e. Everyone should eat more fats because they are good for your skin and hair, and help energize your body.

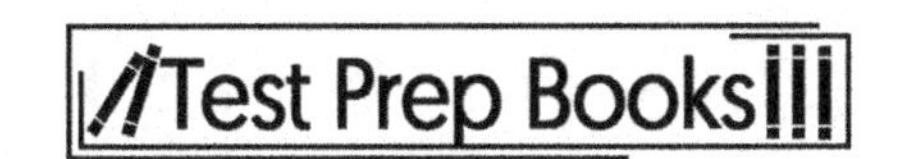

18. John looks like a professional bodybuilder. He weighs 210 pounds and stands six feet tall, which is the size of an NFL linebacker. John looks huge when he enters the room. Years of gym time have clearly paid off in spades.

Which of the following, if true, weakens the argument?

a. John prefers to work out in the morning.
b. The average professional bodybuilder is considerably heavier and taller than the average NFL linebacker.
c. John weighed considerably less before he started working out.
d. John's father, brothers, and male cousins all look like professional bodybuilders, and none of them have ever worked out.
e. John works out five times every week.

19. Hank is a professional writer. He submits regular columns at two blogs and self-publishes romance novels. Hank recently signed with an agent based in New York. To date, Hank has never made any money off his writing.

The strength of the argument depends on which of the following?

a. Hank's agent works at the biggest firm in New York.
b. Being a professional writer requires representation by an agent.
c. Hank's self-published novels and blogs have received generally positive reviews.
d. Being a professional writer does not require earning money.
e. Hank writes ten thousand words per day.

20. Quillium is the most popular blood pressure regulating prescription drug on the market. Giant Pharma, Inc., the largest prescription drug manufacturer in the country, owns the patent on Quillium. Giant Pharma stock is hitting unprecedented high valuations. As a result, Quillium is by far the most effective drug available in treating irregular blood pressure.

Which of the following, if true, most weakens the argument?

a. The most lucrative and popular pharmaceuticals are not always the most effective.
b. Quillium passed the FDA drug testing and screening faster than any other drug.
c. Giant Pharma gouges its customers on Quillium's price.
d. Giant Pharma's high stock prices are attributable to recent patent acquisitions other than Quillium.
e. Quillium has numerous alternate applications.

21. Julia joined Michael Scott Paperless Company, a small New York based tech start-up company, last month. Michael Scott Paperless recently received a valuation of ten million dollars. Julia is clearly the reason for the valuation.

Which of the following statements, if true, most weakens the argument?

a. Michael Scott Paperless Company released an extremely popular mobile application shortly before hiring Julia.
b. Michael Scott Paperless Company is wildly overvalued.
c. Julia is an expert in her field.
d. Julia only started working two weeks before the valuation.
e. Julia completed two important projects during her first month with the company.

22. Advertisement: Cigarettes are deadly. Hundreds of thousands of people die every year from smoking-related causes, such as lung cancer or heart disease. The science is clear—smoking a pack per day for years will shorten one's life. Sitting in a room where someone is smoking might as well be a gas chamber in terms of damage to long-term health.

Which one of the following best describes the flaw in the author's reasoning?

a. The advertisement confuses cause and effect.
b. The advertisement uses overly broad generalization.
c. The advertisement draws an unjustified analogy.
d. The advertisement relies on shoddy science.
e. The advertisement makes an unreasonable logical leap.

23. Jake works for Bank Conglomerate of America (BCA), the largest investment bank in the United States. Jake has worked at Bank Conglomerate of America for a decade. Every American investment bank employs dozens of lawyers to defend against insider-trading allegations. Some Bank Conglomerate of America employees must pass a certification course. However, all employees must complete a mandatory class on insider trading.

If the statements above are correct, which of the following must not be true?

a. Jake took a class on insider trading.
b. Jake passed a certification course.
c. Jake has worked at Bank Conglomerate of America for a decade.
d. Jake never took a class on insider trading.
e. No investment bank has ever been formally charged with insider trading.

24. Leslie lost her job as a cashier at Locally Sourced Food Market because the store went out of business. Two days later, Randy's Ammunition Warehouse closed down for good in the same shopping center. Therefore, the Locally Sourced Food Market's closing clearly caused Randy's to close.

The flawed reasoning in which of the following arguments most mirrors the flawed reasoning presented in the argument above:

a. The United States fought two wars while cutting taxes. The budget deficit continued to increase during that time, which increased national debt. Therefore, fighting two wars and cutting taxes clearly caused an increase in national debt.
b. Tito's Taco Shop recently closed down due to lack of foot traffic. Nearby Bubba's Burrito Bowls also closed down later that month for the same reason. Therefore, a lack of foot traffic caused both businesses to close.
c. Angela recently ran into some rotten luck. Last week she fell off her skateboard, and two days later, she crashed her car. Therefore, Angela needs to recover from her injuries.
d. Theresa lost her job on Monday, but she received an unsolicited offer to consult for a hedge fund that same day. Therefore, losing one job led to another one.
e. Tammy overslept and missed her early class. That same day, she experienced car trouble and missed her night class. Therefore, Tammy did not go to school today.

25. The assassination of Archduke Franz Ferdinand of Austria is often ascribed as the cause of World War I. However, the assassination merely lit the fuse in a combustible situation since many of the world powers were in complicated and convoluted military alliances. For example, England, France, and Russia entered into a mutual defense treaty seven years prior to World War I. Even without Franz Ferdinand's assassination ______________________.

Which of the following most logically completes the passage?

a. A war between the world powers was extremely likely.
b. World War I never would have happened.
c. England, France, and Russia would have started the war.
d. Austria would have started the war.
e. The world powers would still be in complicated and convoluted military alliances.

Sentence Correction

Select the best version of the underlined part of the sentence. The first choice is the same as the original sentence. If you think the original sentence is best, choose the first answer.

26. An important issues stemming from this meeting is that we won't have enough time to meet all of the objectives.

a. An important issues stemming from this meeting
b. Important issue stemming from this meeting
c. An important issue stemming from this meeting
d. Important issues stemming from this meeting
e. Stems from this meeting important issues

27. The rising popularity of the clean eating movement can be attributed to the fact that experts say added sugars and chemicals in our food are to blame for the obesity epidemic.

a. to the fact that experts say added sugars and chemicals in our food are to blame for the obesity epidemic.
b. in the facts that experts say added sugars and chemicals in our food are to blame for the obesity epidemic.
c. to the fact that experts saying added sugars and chemicals in our food are to blame for the obesity epidemic.
d. with the facts that experts say added sugars and chemicals in our food are to blame for the obesity epidemic.
e. to the fact that experts say added sugars and chemicals in our food to be blamed for the obesity epidemic.

28. Tammy is headed to Vancouver on monday morning for a five-day business trip. She's looking for a suitcase that can fit all of her clothes, shoes, accessories, and makeup.

a. monday morning for a five-day business trip.
b. Monday morning for a five-day business trip.
c. Monday morning for a five day business trip.
d. monday morning for a five day business trip.
e. Monday morning for a five-day-business trip.

29. On "take your son to work day," the team found out that Lily's son, Shawn, started taking guitar lessons while he wanted to become a better musician.
    a. while he wanted to become a better musician.
    b. because he wants to become a better musician.
    c. even though he wanted to become a better musician.
    d. because he wanted to become a better musician.
    e. He wanted to become a better musician.

30. Considering the recent rains we have had, it's a wonder the plants haven't drowned.
    a. Considering the recent rains we have had, it's a wonder
    b. Consider the recent rains we have had, it's a wonder
    c. Considering for how much recent rain we have had, its a wonder
    d. Considering, the recent rains we have had, its a wonder
    e. Considering the recent rains we have had it's a wonder

31. Since none of the furniture were delivered on time, we have to move in at a later date.
    a. Since none of the furniture were delivered on time,
    b. Since none of the furniture was delivered on time
    c. Since all of the furniture were delivered on time,
    d. Since all of the furniture was delivered on time
    e. Since none of the furniture was delivered on time,

32. It is necessary for instructors to offer tutoring to any students who need extra help in the class.
    a. to any students who need extra help in the class.
    b. for any students that need extra help in the class.
    c. with any students who need extra help in the class.
    d. for any students needing any extra help in their class.
    e. by any students who need extra help in the class.

33. When we went back to the office after Christmas, I admitted to Janine that my son was not too keen on the gift he had asked for. The fact the train set only includes four cars and one small track was a big disappointment to my son.
    a. the train set only includes four cars and one small track was a big disappointment
    b. that the trains set only include four cars and one small track was a big disappointment
    c. that the train set only included four cars and one small track was a big disappointment
    d. that the train set only includes four cars and one small track were a big disappointment
    e. the train set only includes four cars and one small track were a big disappointment

34. Because many people feel there are too many distractions to get any work done, I actually enjoy working from home.
    a. Because many people
    b. While many people
    c. Maybe many people
    d. With most people
    e. If many people

35. There were many questions about what causes the case to have gone cold, but the detective wasn't willing to discuss it with reporters.
    a. about what causes the case to have gone cold
    b. about why the case is cold
    c. about what causes the case to go cold
    d. about why the case went cold
    e. about what causing the case to have gone cold

36. She went above and beyond that week while she planned on asking for a raise.
    a. while she planned on asking for a raise.
    b. because she plans on asking for a raise.
    c. if she plans on asking for a raise.
    d. for she planned on asking for a raise.
    e. because she planned on asking for a raise.

37. Human resources was called in to settle the dispute, the situation had escalated beyond management control.
    a. Human resources was called in to settle the dispute,
    b. Human resources was called in to settle the dispute;
    c. Human resources were called in to settle the dispute;
    d. Human resources were called in to settle the dispute,
    e. Human resources was called in to settle the dispute:

38. The boss's plan to implement and encourage a wellness program in order to improve the company's health worked almost instantly.
    a. The boss's plan to implement and encourage a wellness program
    b. The boss plan to implement and encourage a wellness program
    c. The bos's plan to implement and encourage a wellness program
    d. The boss's plan to implementing and encourage a wellness program
    e. The boss's plan to implement and encouraging a wellness program

39. Jeanine the co-CEO of Green Eating, stepped up to become sole CEO after her partner stepped down last year.
    a. Jeanine the co-CEO of Green Eating, stepped up
    b. Jeanine the co-CEO of Green Eating stepped up
    c. Jeanine, the co-CEO of Green Eating, stepped up
    d. Jeanine the co-CEO, of Green Eating, stepped up
    e. Jeanine, the co-CEO of Green Eating stepped up

40. Their productivity increased after an extra thirty minutes was added onto their lunch break.
    a. after an extra thirty minutes was added onto their lunch break.
    b. after an extra thirty minutes is added onto their lunch break.
    c. after an extra thirty minutes is adding onto their lunch break.
    d. after an extra thirty minutes were added onto their lunch break.
    e. after an extra thirty minutes was adding onto their lunch break.

41. Although the new startup was wildly successful, <u>the first few years of carrying the business were unsure by Harry due to the inexperienced team.</u>

a. the first few years of carrying the business were unsure by Harry due to the inexperienced team.
b. Harry was unsure how the inexperienced team was going to carry the business through the first few years.
c. carrying the business the first few years the team was inexperienced and Harry was unsure.
d. the inexperienced team carrying the business for the first few years Harry was unsure of.
e. how the inexperienced team was going to carry the business for the first few years Harry was very unsure.

Answer Explanations #1

Integrated Reasoning

Multi-Source Reasoning

1. YES: All three communications are concentrated on the continuance of the Blueberry Pie ice cream flavor, by the focus on how the flavor is progressing and what future options there are for the flavor, whether a year-round flavor, a seasonal flavor, etc.

NO: There is no statement to elude to or confirm that the Blueberry Pie ice cream flavor was one of the company's original flavors.

NO: While it is stated that the Blueberry Pie flavor is a great seller, it doesn't give an indication as to the exact ranking of the flavor.

2. YES: All given communications focus on the future of the Blueberry Pie ice cream flavor. With the production cost changing, it will clearly be a topic covered in the next meeting.

YES: Both offer suggestions for adapting to the changes in production costs to continue the flavor.

3. YES: July 4 is a designated federal holiday, and an employee who works an 8-hour shift on that date is eligible for 8 hours of holiday pay.

NO: July 4 is a designated federal holiday, but an employee can only earn a maximum of 8 hours of holiday pay per holiday. The additional 2 hours would need to be classified as overtime.

4. E: See breakdown of maximum possible hours of premium pay.

- Sunday: 8 hours of Sunday differential
- Monday: 6 hours of regular pay + 2 hours of night differential
- Tuesday: 8 hours of overtime
- Wednesday: 8 hours of regular pay
- Thursday: 1 hour of night differential + 7 hours of regular pay
- Friday: 3 hours of regular pay + 5 hours of night differential
- Ignoring the hours of regular pay, $8 + 2 + 8 + 1 + 5 = 24$

Graphic Interpretation

5. A: This can be determined by calculating the slope of the line at any point or over the entire graph.

$$\frac{1.00 - 3.00}{9 - 1} = \frac{-2}{8} = \frac{-1}{4} = -0.25$$

The slope is -.25. This means that there is a $.25 decrease in cost per slice for every additional slice demanded.

B: A negative correlation shows a steady decrease in the values of the *y* (vertical) axis as the *x* (horizontal) axis increases. Choice *A* would show a steady increase in the *y*-axis values with a steady increase in the *x*-axis values, and Choice *C* would not show any apparent pattern.

6. B: Since this choice has an irrational portion of $\sqrt{3}$, it would be classified as an irrational number. Choice *A* is a whole number, *C* is a rational fraction or repeating decimal, and *D* is the negative version of *C*.

7. D: While all choices (*A, B, C,* and *D*) contain large spikes in discrepancies between the real house prices and the survey house prices, January 2006 to January 2008 covers the largest window of discrepancies.

C: If you follow the line of survey house prices, it is below the real house price until approximately January of 1989 on the chart.

Two-Part Analysis

8. Time together = 1.875; Difference = 1.125

The method of calculating their combined analysis time is as follows:

$$\frac{1}{5} + \frac{1}{3} = \frac{1}{x}$$

$$\frac{15x}{1}\left(\frac{1}{5} + \frac{1}{3}\right) = \frac{15x}{1}\left(\frac{1}{x}\right)$$

$$3x + 5x = 15$$

$$8x = 15$$

$$x = \frac{15}{8} = 1.875$$

The difference would be calculated as follows:

$$3 - 1.875 = 1.125$$

All other choices are miscalculations of the rate or difference, particularly substituting in the wrong given rates.

9. Employee B's pay = $62{,}000 + (b - 20)(15{,}000) + (b - 10)(0.05)(62{,}000)$

This calculates the base pay plus any bonus over 20 sales plus the bonus for sales over 10.

Pay difference between A and B = $62{,}000 + (b - 20)(15{,}000) + (b - 10)(0.05)(62{,}000) - (54{,}000 + (a - 10)(0.05)(54{,}000))$

Other choices are miscalculations using the wrong adjustment for percentages or utilizing the wrong base salaries.

Table Analysis

10. FALSE: The average number of years is 13; the calculation is as follows:

$$\frac{10 + 27 + 2}{3} = 13$$

TRUE: The calculation is as follows:

Management salary average:

$$\frac{81{,}000 + 92{,}000}{2} = 86{,}500$$

Sales salary average:

$$\frac{65{,}000}{1} = 65{,}000$$

Difference = $86{,}500 - 65{,}000 = 21{,}500$

TRUE: Addition of the total years for each department shows:

- HR = 38
- Warehouse = 39
- Marketing = 17
- Sales = 4
- Payroll = 19
- Management = 15

11. FALSE: A plot of the points would show that there is no real correlation between years of experience and salary.

12. TRUE: Numbers from the chart show boy gamers to girl gamers = 20:16, reduced to 5:4.

FALSE: The calculation of this percent is as follows:

Boy gamers $27 + 23 = 50$

Girl gamers $15 + 23 = 38$

$38 = 50x$

$x = 0.76 = 76\ percent$

TRUE: A plot of the number of girl gamers versus the number of times played per week would show the number increasing as the number of times played per week decreased.

Quantitative Reasoning

1. A: Compare each numeral after the decimal point to figure out which overall number is greatest. In answers *A* (1.43785) and *C* (1.43592), both have the same tenths (4) and hundredths (3). However, the thousandths is greater in answer *A* (7), so *A* has the greatest value overall.

2. D: By grouping the four numbers in the answer into factors of the two numbers of the question (6 and 12), it can be determined that:

$$(3 \times 2) \times (4 \times 3) = 6 \times 12$$

Alternatively, each of the answer choices could be prime factored or multiplied out and compared to the original value. 6×12 has a value of 72 and a prime factorization of $2^3 \times 3^2$. The answer choices respectively have values of 64, 84, 108, 72, and 144 and prime factorizations of 2^6, $2^2 \times 3 \times 7$, $2^2 \times 3^3$, $2^3 \times 3^2$, and $2^4 \times 3^2$, so answer *D* is the correct choice.

3. C: The sum total percentage of a pie chart must equal 100%. Since the CD sales take up less than half of the chart and more than a quarter (25%), it can be determined to be 40% overall. This can also be measured with a protractor. The angle of a circle is 360°. Since 25% of 360° would be 90° and 50% would be 180°, the angle percentage of CD sales falls in between; therefore, it would be Choice *C*.

4. B: Since $850 is the price *after* a 20% discount, $850 represents 80% of the original price. To determine the original price, set up a proportion with the ratio of the sale price (850) to original price (unknown) equal to the ratio of sale percentage (where x represents the unknown original price):

$$\frac{850}{x} = \frac{80}{100}$$

To solve a proportion, cross multiply the numerators and denominators and set the products equal to each other:

$$(850)(100) = (80)(x)$$

Multiplying each side results in the equation $85{,}000 = 80x$.

To solve for x, divide both sides by 80: $\frac{85{,}000}{80} = \frac{80x}{80}$, resulting in $x = 1062.5$. Remember that x represents the original price. Subtracting the sale price from the original price ($\$1062.50 - \$850$) indicates that Frank saved $212.50.

5. E: 85% of a number means that number should be multiplied by 0.85: $0.85 \times 20 = \frac{85}{100} \times \frac{20}{1}$, which can be simplified to:

$$\frac{17}{20} \times \frac{20}{1} = 17$$

6. B: Using the conversion rate, multiply the projected weight loss of 25 lb by 0.45 $\frac{kg}{lb}$ to get the amount in kilograms (11.25 kg).

7. D: First, subtract $1,437 from $2,334.50 to find Johnny's monthly savings; this equals $897.50. Then, multiply this amount by 3 to find out how much he will have in three months before he pays for his vacation: this equals $2,692.50. Finally, subtract the cost of the vacation ($1,750) from this amount to find how much Johnny will have left: $942.50.

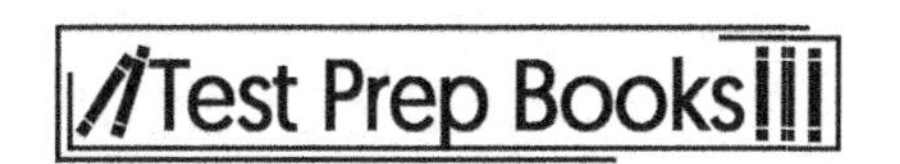

8. B: Dividing by 98 can be approximated by dividing by 100, which would mean shifting the decimal point of the numerator to the left by 2. The result is 4.2, which rounds to 4.

9. D: To find the average of a set of values, add the values together and then divide by the total number of values. In this case, include the unknown value of what Dwayne needs to score on his next test, in order to solve it.

$$\frac{78 + 92 + 83 + 97 + x}{5} = 90$$

Add the unknown value to the new average total, which is 5. Then multiply each side by 5 to simplify the equation, resulting in:

$$78 + 92 + 83 + 97 + x = 450$$
$$350 + x = 450$$
$$x = 100$$

Dwayne would need to get a perfect score of 100 in order to get an average of at least 90.

Test this answer by substituting back into the original formula.

$$\frac{78 + 92 + 83 + 97 + 100}{5} = 90$$

10. D: For an even number of total values, the median is calculated by finding the mean or average of the two middle values once all values have been arranged in ascending order from least to greatest. In this case, $(92 + 83) \div 2$ would equal the median 87.5, Choice *D*.

11. C: Follow the order of operations in order to solve this problem. Solve the parentheses first, and then follow the remainder as usual.

$$(6 \times 4) - 9$$

This equals $24 - 9$ or 15, answer *C*.

12. D: Three girls for every two boys can be expressed as a ratio: 3:2. This can be visualized as splitting the school into 5 groups: 3 girl groups and 2 boy groups. The number of students which are in each group can be found by dividing the total number of students by 5:

$$\frac{650 \text{ students}}{5 \text{ groups}} = \frac{130 \text{ students}}{\text{group}}$$

To find the total number of girls, multiply the number of students per group (130) by the number of girl groups in the school (3). This equals 390, Choice *D*.

13. E: If the average of all six numbers is 6, that means

$$\frac{a + b + c + d + e + x}{6} = 6$$

The sum of the first five numbers is 25, so this equation can be simplified to:

$$\frac{25 + x}{6} = 6$$

Multiplying both sides by 6 gives $25 + x = 36$, and *x*, or the sixth number, can be solved to equal 11.

14. C: Kimberley worked 4.5 hours at the rate of \$10/h and 1 hour at the rate of \$12/h. The problem states that her pay is rounded to the nearest hour, so the 4.5 hours would round up to 5 hours at the rate of \$10/h.

$$(5h)(\$10/h) + (1h)(\$12/h) = \$50 + \$12 = \$62$$

15. C: The first step is to depict each number using decimals. $\frac{91}{100}$ = 0.91

Dividing the numerator by the denominator of $\frac{4}{5}$ to convert it to a decimal yields 0.80, while $\frac{2}{3}$ becomes 0.66 recurring. Rearrange each expression in ascending order, as found in Choice *C*.

16. B: First, calculate the difference between the larger value and the smaller value.

$$378 - 252 = 126$$

This means that there was an increase of 126 customers. Next, find what percent of 252 this would be. To do this set 126 over 252.

$$\frac{126}{252} = .5$$

Now multiply by 100 to turn this into a percent.

$$0.5 \times 100 = 50\%$$

17. A: To calculate the range in a set of data, subtract the highest value with the lowest value. In this graph, the range of Mr. Lennon's students is 5, which can be seen physically in the graph as having the smallest difference compared with the other teachers between the highest value and the lowest value.

18. A: To find the fraction of the bill that the first three people pay, the fractions need to be added, which means finding the common denominator. The common denominator will be 60:

$$\frac{1}{5} + \frac{1}{4} + \frac{1}{3} = \frac{12}{60} + \frac{15}{60} + \frac{20}{60} = \frac{47}{60}$$

The remainder of the bill is:

$$1 - \frac{47}{60} = \frac{60}{60} - \frac{47}{60} = \frac{13}{60}$$

19. B: Simplify each mixed number of the problem into a fraction by multiplying the denominator by the whole number and adding the numerator:

$$\frac{14}{3} - \frac{31}{9}$$

Since the first denominator is a multiple of the second, simplify it further by multiplying both the numerator and denominator of the first expression by 3 so that the denominators of the fractions are equal.

$$\frac{42}{9} - \frac{31}{9} = \frac{11}{9}$$

Simplifying this further, divide the numerator 11 by the denominator 9; this leaves 1 with a remainder of 2. To write this as a mixed number, place the remainder over the denominator, resulting in $1\frac{2}{9}$.

20. A: The total fraction taken up by green and red shirts will be:

$$\frac{1}{3} + \frac{2}{5} = \frac{5}{15} + \frac{6}{15} = \frac{11}{15}$$

The remaining fraction is:

$$1 - \frac{11}{15} = \frac{15}{15} - \frac{11}{15} = \frac{4}{15}$$

21. B: Statement 1 does not give enough information to determine the measure of the angle since only the length of the hypotenuse is given. If the length of either of the other two sides was also given, trigonometric ratios could be used to determine the angle. However, without this information, the angle cannot be determined.

Using statement 2, the value of the angle can be calculated as follows:

$$180 - (90 + 30) = 60$$

22. C: Statement 1 relays the basic tenets necessary to assign measures to angles between two parallel lines traversed by a single line. Statement 2 supplies the information necessary to use the rule for opposite exterior angle measures, thus assigning the value of angle 2 = 53 degrees.

23. C: The information from both statements is necessary to answer the question. The equation in Statement 1 can be factored to find $x = -2$ and $x = 1$.

$$x^2 + x - 2 = 0$$

$$(x + 2)(x - 1) = 0$$

$$x = -2, x = 1$$

Because Statement 2 states that $x > 0$, $x = 1$ can be plugged into the original equation to find the numerical value of that equation.

24. A: Knowing that all three legs are equal is enough to solve the problem. If you let x represent the length of each leg, then the total length of the race becomes:

$$x + x + x = 3x$$

Then the time for each leg is the distance for each leg divided by the speed. Then getting like denominators gives:

$$\frac{3x}{150} + \frac{3x}{150} + \frac{2x}{150} = \frac{8x}{150}$$

Then, the average speed calculation is total distance over total time. So, you end up with:

$$\frac{3x}{\left(\frac{8x}{150}\right)}$$

Then solve from here:

$$3x \times \frac{150}{8x}$$

Cancel the x:

$$3 \times \frac{150}{8}$$

$$\frac{450}{8} = 56.25$$

Statement 2 is not sufficient because knowing the total length does not tell you what distances were traversed at 75 mph or 50 mph, so the time elapsed and thus average speed cannot be solved from the total distance alone.

25. A: To calculate a value, statement 1 must be established to determine the answer is not 0. The information in statement 2 is not necessary to complete the solution.

Calculation is as follows:

$$40\ percent = 0.40$$
$$0.40(20w) = 8w$$

26. A: To solve this, only the information in statement 1 is necessary. The calculation of the solution is as follows:

$a^2 - b^2$ is the difference of two squares, so it can be factored out into:

$$(a - b)(a + b) = 12$$

Substitute the information from statement 1 into the factored problem.

$$4(a + b) = 12$$

Solving for *a* yields $a = 3 - b$.

Plug that back into the original equation:

$$(3 - b)^2 - b^2 = 12$$

$$9 - 6b + b^2 - b^2 = 12$$

$$-6b = 3$$

$$b = -\frac{1}{2}$$

$$a - -\frac{1}{2} = 4$$

$$a = 3.5$$

Then, using the values of *a* and *b* found, *ab* can be determined:

$$ab = 3.5 \times -\frac{1}{2} = -1.75$$

27. E: Neither statement, even when taken together, provide enough information to answer the question. We need two equations to be able to set up simultaneous equations to solve together since we have two variables (the number of cans of red paint and the number of cans of blue paint). With statement 1, we are only told that one of the colors was $1 more than the other, but since we aren't told which is more expensive, we cannot form an equation that puts both colors in terms of the same variable. Similarly, with statement 2, we are only told she bought two more blue cans than red ones. Did she buy 1 red can and 3 blue? 2 red and 4 blue? 3 red and 5 blue? All we know is she bought at least 3 cans since she bought two more blue cans than red cans.

Because we don't even know exactly how much of the $16 she spent (she might have received change back), we can't work backwards and try plugging in numbers to try and see what numbers work out evenly. Moreover, we aren't told that the cans sell for an even dollar amount, just that one is $1 more than another. Perhaps one color is $1.50 and then the other is $2.50. Likewise, it's possible that one color is $2.10 and the other is $3.10. The possibilities are vast. There is not enough information to solve the problem.

28. C: Both statements are necessary for completing the calculations as follows:

If $x > 0$, then 5y will also be > 0.

Factor the equation into $\sqrt{(x - y)^2} = 16$.

Then, take the square root of both sides and substitute $x = 5y$.

$$5y - y = 4$$

Therefore, $4y = 4$, so $y = 1$. Because the second statement says that $x = 5y$, $x = 5$.

29. B: Statement 1 is not necessary information, but with statement 2, the calculation can be completed as follows:

$$10 + b = 9 \qquad\qquad 10 - c = 14$$
$$b = -1 \qquad\qquad c = -4$$

$$c - b = -4 - -1$$
$$-3$$

30. A: Statement 1 is the only information necessary to complete the calculation as follows:

$$12\ percent = 0.12$$
$$156 = 0.12x$$
$$x = \$1{,}300$$

31. B: The information included in statement 1 is obtainable by inspection of the given figure and therefore redundant. Statement 2 provides the information necessary to complete the calculation as follows:

$$KL + LM = 16 \qquad\qquad KL = 16 - LM$$

$$LM + MN = 20 \qquad\qquad MN = 20 - LM$$

$$KN = KL + MN + LM = 30$$

$$16 - LM + 20 - LM + LM = 30$$

$$36 - 30 = LM$$

$$6 = LM$$

32. D: Each statement alone could be used to answer the question.

This is a right triangle; therefore, trigonometry can be used to find the length of the missing side.

Using statement 1:

$$\sin(53.1 \text{ degrees}) = \frac{x}{50\,feet}$$

$$0.7997 = \frac{x}{50\,feet}$$

$$x = 39.985\,feet$$

The ladder reaches approximately 40 feet up the wall.

Using statement 2:

$$\cos(36.9 \text{ degrees}) = \frac{x}{50 \text{ feet}}$$

$$0.7997 = \frac{x}{50 \text{ feet}}$$

$$x = 39.985 \text{ feet}$$

The same answer was obtained. The ladder reaches approximately 40 feet up the wall.

33. A: Only statement 1 has information necessary to complete the calculations as follows:

$$\frac{2}{5}(804) = 321.6$$

$$\frac{1}{4}(321.6) = 80.4$$

$$80 \text{ positions}$$

34. D: The information in either statement 1 or statement 2 can be used to solve the problem with the calculations as follows:

$$L = 2W$$
$$Area = L \times W = 2W \times W = 2W^2$$
$$A = 2(2)^2 = 8\ units$$

$$OR$$

$$2L + 2W = 12$$
$$2(2W) + 2W = 12$$
$$4W + 2W = 12$$
$$6W = 12$$
$$W = 2$$

$$Area = L \times W = 2W \times W = 2W^2$$
$$2(2)^2 = 8\ units$$

35. B: Only Statement 2 is necessary to calculate the solution as follows:

It's not necessary to know how many marbles there are in total since only one red marble is drawn from the bag of three colors. Statement 2 says that there is a $\frac{1}{2}$ chance of drawing a white marble and a $\frac{1}{4}$ chance of drawing a blue marble. Therefore, there is a $\frac{1}{4}$ chance of drawing a red marble since the sum of these fractions must equal 1.

36. B: Statement 1 contains frivolous information, but statement 2 contains information necessary to complete the calculation as follows:

3 meats × 3 cheeses × 5 breads = 45 choices for sandwiches

37. E: Both statements together do not provide enough information to solve the problem. We are looking to find the perimeter of each of the two plots, then add these values together in linear feet, to determine how much fencing the farmer needs. We are provided with the area of each plot but told nothing about the shape or length of any of the sides. Are they squares? Long, thin rectangles? Circles? The perimeter will be different in either condition.

Consider, for example, the plot in statement 2, which has an area of 10 square feet. Imagining possibilities for rectangle plots (though it could be another shape), area is length × width. Theoretically, the plot could be 1 foot wide and 10 feet long and yield an area of 10 feet or it could be 2.5 feet wide and 4 feet long and still yield an area of 10 feet. However, in just these two cases alone, the perimeter would be different. Recall that the formula for a perimeter of a rectangle is:

$$\text{perimeter} = 2l + 2w$$

Therefore, in the first case, the perimeter would be $(2 \times 10) + (2 \times 1)$, which equals 22 feet. In the second condition, the perimeter would be $(2 \times 4) + (2 \times 2.5)$, which only equals 13 feet. This plot shape would require 9 fewer feet of fencing. A similar conundrum exists for the larger plot. From this example, it can be seen that there is simply not enough information to conclusively solve the problem.

Verbal Reasoning

Reading Comprehension

1. C: Choice *A* is incorrect because the seaport is noted as "thriving"; also, the slave trading companies were noted as being "lucrative." Choice *B* is incorrect because the slave traders actually built both office structures and slave holding buildings in downtown Alexandria; there is no mention of renting, or of landlords. Choice *C* is correct because Joseph Bruin bought hundreds of slaves during the years 1844 to 1861. Choice *D* is incorrect because the Bruin Slave Jail is stated as NOT open to the public. Choice *E* is incorrect because the passage notes that the offices and slave holding units were open until the end of the Civil War.

2. A: Choices *B, C, D,* and *E* can all be regarded as false based on the information provided in the passage. Choice *A* contains information provided in the passage; therefore, the statement is true. Choice *B* is false because the passage infers that Joseph Bruin was notorious as a slave trader; in fact, two sisters tried to run away from Joseph Bruin. Choice *C* is false because the slave pen was not four stories high; the passage specifically noted that the slave pen was two stories high. Choice *D* is false because the passage does not refer to Union or Confederate soldiers, and the Bruin Slave Jail was what became the Fairfax County courthouse. Choice *E* is false because there is no information in the passage that indicates that literature, like *Uncle Tom's Cabin*, was preserved by the Northern Virginia Urban League.

3. D: Choice *A* is false, based on the passage statement that Alexandria was founded in 1749. Choice *B* is false because the passage does not suggest that Harriet Beecher Stowe was a slave; rather, the passage states that Stowe was the author of *Uncle Tom's Cabin.* Choice *C* is incorrect; the passage claims that northern abolitionists tried to save two Christian slave sisters from a fate of prostitution. Choice *D* is true based on the information found in the passage. Choice *E* is false; the town of Annapolis is not cited in the passage.

4. C: The purpose of the passage is to shed light on the history of Joseph Bruin's Slave Jail and what became of it. Choice *A* is incorrect because while the two sisters are mentioned in the story to provide details, they are not the main purpose of the story. Choice *B* is incorrect because while the beginning of the story contains the information about the town and its slave business, this answer option leaves out the fact that the passage is focused on one slave jail in particular and omits anything about the conclusion of the passage, which is actually key in the main focus of the passage—how Joseph Bruin's Slave Jail came about and what became of it. Choice *D* is incorrect because the point of the passage is not about where the historical Bruin Slave Jail currently stands but the history behind it.

5. C: Because the details in Choice *A* and Choice *B* are examples of how an emotionally intelligent leader operates, they are not the best choice for the definition of the term *emotional intelligence*. They are qualities observed in an EI leader. Choice *C* is true as noted in the second sentence of the passage: Emotional Intelligence (EI) includes developing the ability to know one's own emotions, to regulate impulses and emotions, and to use interpersonal communication skills with ease while dealing with other people. It makes sense that someone with well-developed emotional intelligence will have a good handle on understanding their emotions, be able to regulate impulses and emotions, and use interpersonal communication skills. Choice *D* is not a definition of EI. Choice *E* is the opposite of the definition of EI, so both Choice *D* and Choice *E* are incorrect.

6. C: Choice *E* can be eliminated immediately because of the signal word "obviously." Choice *A* can be eliminated because it does not reflect an accurate fact. Choices *B* and *D* do not support claims about how to be a successful leader.

7. E: The qualities of an unsuccessful leader possessing a transactional leadership style are listed in the passage. Choices *A* and *B* are incorrect because these options reflect the qualities of a successful leader. Choices *C* and *D* are definitely not characteristics of a successful leader; however, they are not presented in the passage and readers should do their best to ignore such options.

8. D: Even though some choices may be true of successful leaders, the best answer must be supported by sub-points in the passage. Therefore, Choices *A* and *C* are incorrect. Choice *B* is incorrect because uncompromising transactional leadership styles squelch success. Choice *E* is never mentioned in the passage.

9. E: To support Choice *E*, the idea that a leader can develop emotional intelligence if desired, the passage says, "There are ways to develop emotional intelligence for the person who wants to improve their leadership style." After a careful reading of the passage about emotional intelligence, readers can select points for the statements that are contradictory in the selection. For example, the statement that contradicts Choice *A* says, "Becoming a successful leader in today's industry, government, and nonprofit sectors requires more than a high intelligence quotient (IQ)." Likewise, a contradictory passage for Choice *C* is: "Building relationships outside the institution with leadership coaches and with professional development trainers can also help leaders who want to grow their leadership success." Choices *B* and *D* do not have supporting evidence in the passage to make them true.

10. B: Readers should carefully focus their attention on the beginning of the passage to answer this series of questions. Even though the sentences may be worded a bit differently, all but one statement is true. It presents a false idea that descriptive details are not necessary until the climax of the story. Even if one does not read the passage, he or she probably knows that all good writing begins with descriptive details to develop the main theme the writer intends for the narrative.

11. C: This choice allows room for the fact that not all people who attempt to write a play will find it easy. If the writer follows the basic principles of narrative writing described in the passage, however, writing a play does not have to be an excruciating experience. None of the other options can be supported by points from the passage.

12. C: Choice *C* is true based on the information that claims an image is like using a thousand words. Choice *A* is incorrect based on the sentence that reads, "Next, use dialogue to reveal the attitudes and personalities of each of the characters who have a key part in the unfolding story." Choice *B* is false because drama does not necessarily need to be predictable. Choice *D* contradicts the point that the protagonist should experience self-discovery. Finally, Choice *E* is incorrect because all drama suggests some challenge for the characters to experience.

13. B: To suggest that a ten-minute play is accessible does not imply any timeline, nor does the passage mention how long a playwright spends with revisions and rewrites. So, Choice *A* is incorrect. Choice *B* is correct because of the opening statement that reads, "Learning how to write a ten-minute play may seem like a monumental task at first; but, if you follow a simple creative writing strategy, similar to writing a narrative story, you will be able to write a successful drama." None of the remaining choices are supported by points in the passage.

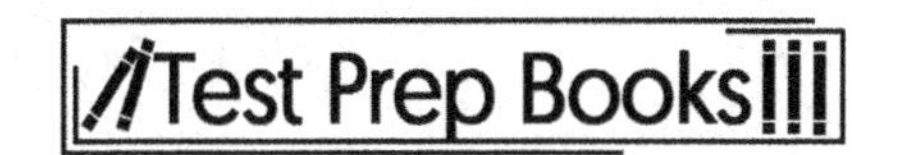

14. E: Note that the only element not mentioned in the passage is the style feature that is part of a narrative writer's tool kit. It is not to say that ten-minute plays do not have style. The correct answer denotes only that the element of style was not illustrated in this particular passage.

Critical Reasoning

15. E: Choice *A* is incorrect. Alexis is not reinforcing Dalton's argument here because she offers no solidarity with his argument that agribusiness produces greenhouse emissions. Therefore, Choice *A* is incorrect.

Choice *B* is also incorrect. Alexis does not add to Dalton's argument in any way, because the foundation of the two arguments are fundamentally different.

Choice *C* is incorrect. It might be tempting to pick Choice *C*; however, it doesn't give us the whole picture. Alexis' response mostly gives an alternative viewpoint.

Choice *D* is incorrect. Similar to Choice *B*, Alexis does not offer any claims to back up Dalton's argument, though she does offer a claim to back up a differing argument.

Choice *E* is the best answer because Alexis is suggesting an alternative factor to the problem of global warming by offering a differing cause other than agribusiness, coupled with a differing solution.

Therefore, Choice *E* is the correct answer.

16. D: Choice *A* is incorrect. There is no evidence that the speaker thinks that Shelby is an excellent coder. They only indicate that she is "proficient." Eliminate this choice.

Choice *B* is incorrect. Shelby has fifteen years of experience, which serves as the speaker's evidence that she is a proficient coder, not that the company is looking for someone with fifteen years of experience.

Choice *C* is also incorrect. If anything, the speaker assumes that Shelby will be a good fit for the company, not a bad fit for the company. The speaker is recommending Shelby for the job.

Choice *D* is the best choice here. The speaker assumes that someone having fifteen years of experience will make them a proficient coder. Although this is quite possibly true, the speaker is still making an assumption about Shelby's skills.

Choice *E* is incorrect. The speaker has not compared Shelby to anyone else in their statement.

Therefore, Choice *D* is the correct answer.

17. B: Choice *A* is incorrect because it offers continued advice for those who agree with the statement.

Choice *B* is the correct answer choice. The speaker's argument says that everyone should eat meat because the only way to fulfill protein needs is from the complete protein in meat. However, if the statement were introduced into the argument that complete protein could be created in an alternative way than simply eating meat, it would weaken the speaker's original argument. Choice *B* is the best choice.

Choice *C* is incorrect. Facts about amino acids would not weaken the presented argument. This information acts as background context for the presented argument.

Choice *D* is misleading. The answer choice states that fruits and vegetables contain essential nutrients that help our bodies fight disease. This doesn't strengthen or weaken the original argument; it's a distractor.

Choice *E* is also a distractor, like Choice *D*. It doesn't strengthen or weaken the original argument, it just provides different information about nutrients.

Therefore, Choice *B* is the correct answer.

18. D: Choice *A* is irrelevant. The argument makes no mention as to when John works out. Would it weaken the conclusion—which is that years of gym time have clearly paid off—if he works out in the morning instead of the afternoon? No, of course not. Eliminate this choice.

Choice *B* preys on those who incorrectly identify the conclusion. Test takers who identify the first sentence as the conclusion will find this answer very appealing. If John is the size of an NFL linebacker, but linebackers are much smaller than professional bodybuilders, then John doesn't look like a professional linebacker. However, Choice *B* is irrelevant as to whether years of working out have paid off. Eliminate this choice.

Choice *C* actually strengthens the argument. If John weighed considerably less before working out and now he looks like a professional bodybuilder, then years of working out have definitely paid off. Eliminate this choice.

Choice *D* looks very appealing. If John's family members are all similar in size without weightlifting, then it's possible that it doesn't matter that John regularly spends time in the gym. Even without lifting, John would likely be the same size as his male family members. Therefore, years of working out would not be the reason why he looks like a professional bodybuilder. Don't be concerned that Choice *D* is unlikely in reality. If a question says something's true, then treat it as true. Keep this choice for now.

Choice *E* reinforces the argument's conclusion. The argument already states that John has gone to the gym for years. Whether he goes three, five, or seven times per week does not weaken the argument. Eliminate this choice.

Therefore, Choice *D* is the correct answer.

19. D: Choice *A* is irrelevant. The argument's conclusion is that Hank is a professional writer. The argument does not depend on whether Hank's agent is the best or worst in the business. Eliminate this choice.

Choice *B* seems fairly strong at first glance. It feels reasonable to say that being a professional writer requires representation. However, the argument would still be strong if being a professional writer did not require an agent. Hank would still be a professional writer. Eliminate this choice.

Choice *C* is irrelevant. Whether Hank is a professional writer does not depend on his reviews. Eliminate this choice.

Choice *D* is strong. Negate it to determine if the argument falls apart. If being a professional writer requires earning money, then Hank would not be a professional writer. The argument falls apart. This is almost definitely the correct answer.

Choice *E* is irrelevant. The argument does not attempt to tie Hank's professionalism with a word count. For the purposes of this argument, it does not matter if Hank writes ten or ten thousand words per day. Eliminate this choice.

Therefore, Choice *D* is the correct answer.

20. A: Choice *A* looks very strong. The argument devotes most of its time discussing Quillium's popularity and monetary value. It uses these facts to conclude that Quillium is the most effective drug in treating blood pressure. If the most lucrative and popular drugs are not necessarily the most effective, then it seriously weakens the argument. Leave this choice for now, and look at the other answer choices.

Choice *B* strengthens the argument. Moving through the screening process at record time supports the conclusion. It definitely does not weaken it. Eliminate this choice.

Choice *C* is irrelevant. Whether Giant Pharma gouges its customers does not affect the conclusion concerning the drug's effectiveness. Eliminate this choice.

Choice *D* is misleading. This choice would greatly weaken if the argument's conclusion were that Quillium is the reason for Giant Pharma's high valuation. However, this is only a premise. Choice *D* weakens this premise, but it does not weaken the argument as much as Choice *A*, which attacks the heart of the argument. Eliminate this choice.

Choice *E* is irrelevant. The argument's conclusion is that Quillium is the most effective drug for treating irregular blood pressure. Does it matter if Quillium has alternate applications? Of course not, so eliminate this choice.

Therefore, Choice *A* is the correct answer.

21. A: Choice *A* is very strong since it provides an alternate explanation for the high valuation other than Julia. If the Company released an extremely popular application, then the application is the real reason for the 10 million dollar valuation. Furthermore, this answer choice explicitly states that the application was released before Julia's hiring. Keep this choice for now.

Choice *B* is irrelevant. Whether investors are properly evaluating the Company's price does not affect Julia's role in that valuation. Eliminate this choice.

Choice *C* strengthens the argument. If Julia is an expert in her field, then her skills could have been the reason for the valuation. Investors could have factored in Julia's expertise in their valuation. It definitely does not weaken the argument. Eliminate this choice.

Choice *D* is another strong answer choice. If Julia only worked at Michael Scott Paperless Company for two weeks, then it is less likely that she's the reason for the 10 million dollar valuation. However, if she's a renowned expert or extreme talent, then her hiring alone could have affected the valuation. This answer choice is less strong than Choice *A*, which provides a clear alternative explanation for the sudden increase in valuation. Since Choice *A* is stronger, eliminate Choice *D*.

Choice *E* strengthens the argument. If Julia completed two important projects during her first month, then she could very well be the reason for the valuation. It definitely does not weaken the argument; eliminate Choice *E* as well.

Therefore, Choice *A* is the correct answer.

22. C: Choice *A* does not identify a flaw in the advertisement's reasoning. The advertisement connects smoking with fatal disease. At no point does the advertisement confuse the cause and effect. Eliminate this choice.

Choice *B* is incorrect. The advertisement does not make any overly broad generalizations. Eliminate this choice.

Choice *C* correctly identifies the argument's flaw. The argument analogizes second hand smoke with a gas chamber without offering any evidence concerning second hand smoke's health risk. The advertisement is clearly relying on hyperbole. The advertisement's argument properly justifies smoking with adverse health effects, but it does not do the same for second-hand smoke. This is most likely the correct answer.

Choice *D* is incorrect. Nothing in the argument states that there's real dispute over smoking's effect on health. Eliminate this choice.

Choice *E* is not present in the argument. Eliminate this choice.

Therefore, Choice *C* is the correct answer.

23. D: Choice *A* is true. According to the argument, all of the employees must complete a mandatory class on insider trading. Jake is an employee. Therefore, he must have taken a class on insider trading. Eliminate this choice.

Choice *B* is not necessarily true; however, it could be true. According to the argument, some employees must pass a certification course, but it does not mention whether Jake is one of those employees. This choice may or may not be true, so it cannot be the correct answer. Eliminate this choice.

Choice *C* restates a premise, so it is true. Therefore, it is incorrect for the purposes of this question. Eliminate this choice.

Choice *D* must be incorrect according to the argument. As previously discussed, all of the employees must take a class on insider trading. Jake is an employee, so he must have taken the class. Therefore, Choice *D* must not be true.

Choice *E* could be true. The argument states that investment banks hire dozens of lawyers. It does not mention whether any investment bank has ever been charged. Since this choice could be true, it's not the correct answer. Eliminate this choice.

Therefore, Choice *D* is the correct answer.

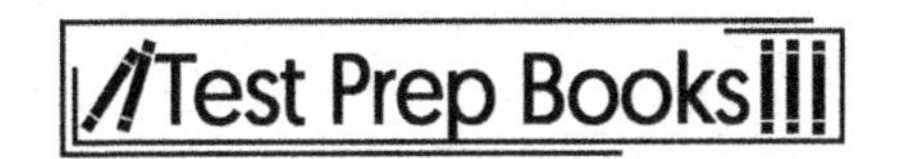

24. D: Choice *A* is incorrect, because it follows logically. This answer choice tells us that the United States increased spending while cutting taxes, which increased debt. This cannot be the correct answer since there's no flawed reasoning.

Choice *B* is similar to the argument since both the argument and answer choice involve a nearby business closing down. However, Choice *B* states that both businesses closed for the same reason. It does not claim that one closing caused the other like the flawed reasoning present in the argument. Rather, it claims that both businesses closed as a result of a single factor. Eliminate this choice.

Choice *C* does not rely on flawed reasoning. It states that Angela fell off her skateboard and crashed her car so she needs to recover from her injuries. This answer choice does not confuse causation with correlation like the argument. Eliminate this choice.

Choice *D* looks promising. The answer choice claims that losing her job caused her to receive another job offer on the same day. This mixes causation with correlation. The argument claims that Locally Sourced Food Market's caused Randy's Ammunition Warehouse to close without offering any evidence aside from time and location. Both share the same flaw—confusing correlation and causation—so it is the correct answer.

Choice *E* is incorrect since it's logically sound. There is no flaw in the choice's reasoning, so it cannot be the correct answer. Eliminate this choice.

Therefore, Choice *D* is the correct answer.

25. A: Choice *A* is consistent with the argument's logic. The argument asserts that the world powers' military alliances amounted to a lit fuse, and the assassination merely lit it. The main point of the argument is that any event involving the military alliances would have led to a world war. This is a very strong answer.

Choice *B* runs counter to the argument's tone and reasoning. It can immediately be eliminated.

Choice *C* is also clearly incorrect. At no point does the argument blame any single or group of countries for starting World War I. This can also be immediately eliminated.

Choice *D* is wrong for the same reason as Choice *C*. Eliminate this choice.

Choice *E* is a better option than the previous three, but it fails to complete the passage in any meaningful way. The argument is contending that the assassination was a sufficient cause for the war, rather than a necessary cause. Choice *A* more logically completes the passage. Eliminate this choice.

Therefore, Choice *A* is correct.

Sentence Correction

26. C: In this answer, the article and subject agree, and the subject and predicate agree. Choice *A* is incorrect because the article (*an*) and the noun (*issues*) do not agree in number. Choice *B* is incorrect because an article is needed before *important issue.* Choice *D* is incorrect because the plural subject *issues* does not agree with the singular verb *is*. Choice *E* is awkward and does not make grammatical sense.

27. A: Choices *B* and *D* both use the expression *attributed to the fact* incorrectly. It can only be attributed *to* the fact, not *with* or *in* the fact. Choice *C* incorrectly uses a gerund, *saying*, when it should use the present tense of the verb *say*. Choice *E* is incorrect because it uses the infinitive "to be blamed" instead of the correct verb usage, "are to blame."

28. B: Choice *B* is correct because it correctly capitalizes "Monday," one of the days of the week. It also puts a hyphen on the compound adjective "five-day." Compound adjectives are used when two adjectives are joined together to modify the same noun, in this case, "business trip."

29. D: In a cause/effect relationship, it is correct to use the word because in the clausal part of the sentence. This can eliminate both Choices *A and* C which don't clearly show the cause/effect relationship. Choice *B* is incorrect because it uses the present tense, when the first part of the sentence is in the past tense. It makes grammatical sense for both parts of the sentence to be in present tense. Choice *E* is technically correct, but it's not the best answer. We are still not given a cause/effect relationship; we are simply splitting the two ideas up when they should be conjoined.

30. A: In Choice *B*, the present tense form of the verb *consider* creates an independent clause joined to another independent clause with only a comma, which is a comma splice and grammatically incorrect. Both *C* and *D* use the possessive form of *its*, when it should be the contraction *it's* for *it is.* Choice *D* also includes incorrect comma placement. Choice *E* needs a comma after the phrase "Considering the recent rains we have had."

31. E: Choice *A* uses the plural form of the verb, when the subject is the pronoun *none*, which needs a singular verb. Choice *B* needs a comma after the dependent clause. Choice *C* also uses the wrong verb form and uses the word *all* in place of *none*, which doesn't make sense in the context of the sentence. Choice *D* uses *all* again, and is missing the comma, which is necessary to set the dependent clause off from the independent clause.

32. A: Answer Choice *A* uses the best, most concise word choice. Choice *B* uses the pronoun *that* to refer to people instead of *who. C* incorrectly uses the preposition *with*. Choice *D* uses the preposition *for* and the additional word *any*, making the sentence wordy and less clear. Choice *E* is incorrect because it uses the preposition *by* instead of *to*.

33. C: Choice *C* is the best answer choice because it keeps the past tense consistent with the verbs "included" and "was." Choice *A* is missing the word *that*, which is necessary for the sentence to make sense. Choice *B* pluralizes *trains* and uses the singular form of the word *include*, so it does not agree with the word *set*. Choice *D* changes the verb to *were*, which is in plural form and does not agree with the singular subject. Choice *E* is missing the word *that* and also should use the word *was* instead of *were*.

34. B: Choice *B* uses the best choice of words to create a subordinate and independent clause. In Choice *A*, *because* makes it seem like this is the reason they enjoy working from home, which is incorrect. In *C*, the word *maybe* creates two independent clauses, which are not joined properly with a comma. Choice *D* uses *with*, which does not make grammatical sense. Choice *E* is incorrect; by providing *if* there is an expected *then* statement that is not represented in the sentence.

35. D: Choices *A* and *C* use additional words and phrases that are not necessary. Choice *B* is more concise, but uses the present tense of *is*. This does not agree with the rest of the sentence, which uses past tense. Choice *E* is incorrect because it is missing the *is* in *is causing*, and is not the best choice. The best choice is Choice *D*, which uses the most concise sentence structure and is grammatically correct.

36. E: Choice *A* is incorrect because *while* does not make sense in the sentence. Choices *B* and *C* use the present tense *plans* instead of *planned*. Choice *D* uses *for* instead of *because* which changes the tone awkwardly. Choice *E*, which uses *because*, makes the most sense because the sentence involves cause and effect.

37. B: Choice *A* is incorrect because a comma in the middle of two independent clauses is considered a comma splice. Choices *C* and *D* are also incorrect; human resources is considered a singular collective noun, so *was* would be used instead of *were*. Choice *E* is incorrect because it ends with a colon, which is usually used for listing things. Choice *B* is the best choice here, because semicolons act to separate two independent clauses that are closely related.

38. A: There are two problems across all the answer choices except for *A*. The word *boss* must be possessive—it is the plan of the boss. Therefore, it must be the *boss's plan*. Secondly, there must be parallel structure with the words *implement* and *encourage*, making Choices *D* and *E* incorrect.

39. C: Choice *C* is correct because the modifying phrase that describes Jeanine, *the co-CEO of Green Eating*, should be marked off with commas.

40. D: The best choice is *D*. The plural noun *thirty minutes* calls for the plural verb *were*. Choices *B* and *C* are incorrect because the verb *is* denotes present tense instead of past tense.

41. B: This is the best answer choice because we have a clear, directly stated independent clause that begins with the subject "Harry" and follows through logically. The other choices are jumbled up clauses that do not take clarity into consideration.

Practice Test #2

AWA Prompt

The following is a persuasive paragraph written for the beginning of a history lecture:

> "Evidently, our country has overlooked both the importance of learning history and the appreciation of it. But why is this a huge problem? Other than historians, who really cares how the War of 1812 began, or who Alexander the Great's tutor was? Well, not many, as it turns out. So, *is* history really that important? Yes! History is *critical* to help us understand the underlying forces that shape decisive events, to prevent us from making the same mistakes twice, and to give us context for current events."

In your essay response, evaluate the underlying logic of the given argument above. Analyze its use of evidence as well as its assumptions. You might also think about what kind of evidence might weaken or strengthen the argument. The essay should be approximately 400–600 words.

Integrated Reasoning

Multi-Source Reasoning

Prompt 1

Currently, the only health insurance plan offered to a certain company's employees is a zero-deductible plan, and its monthly premium is expensive for the employees. A zero deductible plan means that employees do not have to pay a minimum balance before insurance contributes to health care expenses. Currently, the insurance plan pays 90% of all covered medical costs. However, this plan comes at a high cost to the employees, and there will be more options given in open enrollment this year.

Email Announcement: We are changing our company health insurance plan next year. Instead of the flat monthly fee of $500 for 90% coverage, there will be 4 different options for single employees. Details about family plans will be made available next week. The 4 options for single employees are given in the following excerpts:

Excerpt 1

Single employees will keep the same level of coverage (90% of covered medical costs), and instead of paying $500 monthly for health insurance, they will have $250 taken out of each biweekly paycheck.

Excerpt 2

A new option is a high deductible plan. With this plan, employees will have $50 taken out of each biweekly paycheck. Single employees will have a $5,000 deductible. After $5,000 is spent on medical costs, insurance will cover 90% of covered medical costs.

Excerpt 3

A second new option is a different deductible. With this plan, single employees will have $100 taken out of each paycheck. Employees selecting this option will have a $3,000 deductible. After the deductible is spent on medical costs, insurance will cover 85% of covered medical costs.

Excerpt 4

A final option involves a low deductible option. This plan involves $200 taken out of each paycheck, and single employees selecting this option will have a $1,000 deductible. After this amount is spent on medical costs, insurance will cover 80% of covered medical costs.

1. Based on the information provided, answer the following questions.

YES	NO	
O	O	The employees who select the same plan (with no deductible) will be paying the same yearly amount.
O	O	The high deductible plan will be selected the most often because it is the least expensive.
O	O	The low deductible plan will cost each single employee $5,200 a year.

2. Based on the information provided, answer the following questions.

YES	NO	
O	O	An employee who estimates that she will have $10,000 in total covered medical costs next year should choose the high deductible plan.
O	O	An employee who anticipates no medical bills should choose the high deductible plan.
O	O	An employee who estimates that he will have $100,000 in total covered medical costs next year should choose the high deductible plan.

Prompt 2:

Announcement: We are excited to announce that we are overhauling our tuition costs for the 2019–2020 school year. Instead of paying a flat rate per credit hour, student who take more credit hours will have discounts applied. Currently, in-state students pay $650 per credit hour and out-of-state students pay $1,050 per credit hour. These costs will change in the following manner:

Students taking between 1 and 6 credit hours will pay the same amount per credit hour, with a 1% increase next year.

Students taking between 7 and 10 credit hours will pay the same 1% increase per credit hour, but they will receive a 10% discount on their tuition.

Students taking between 11 and 15 credit hours will also see a 1% increase per credit hour, but will receive a 15% discount on tuition.

Students taking between 16 and 18 credit hours will pay a 1% increase per credit hour, and they will receive a 20% discount on total tuition.

Students taking more than 18 credit hours receive the 20% discount for the first 18 credit hours of tuition; every credit hour taken after 18 is subject to the cost of taking a single credit hour.

We are also reworking the enrollment process. Students who plan to take more than 15 credit hours will select classes on the first Monday in April. Students who plan to take between 11 and 15 credit hours will select classes on the second Monday in April. Finally, students who plan to take fewer than 11 credit hours will select classes on the third Monday in April.

We hope that these changes encourage students to take more credit hours each semester in order to graduate sooner.

3. Calculate the cost of taking 12 credit hours as an out-of-state resident in the 2019–2020 school year. Round to the nearest dollar.

a. $10,817
b. $12,726
c. $12,600
d. $10,710
e. $1,909

4. Based on the information provided, answer the following questions.

YES	NO	
O	O	A student registering for 4 credit hours next year will enroll after a student registering for 9 credit hours.
O	O	A student who takes the same number of credit hours as in the previous year will pay a 1% increase in tuition next year.
O	O	An in-state student taking 19 credit hours next year will pay $10,110.10.

Graphic Interpretation

5. The table below shows the number of employees at XYZ Cable Company and their corresponding years of service at the company. What percent of total employees have worked at XYZ Cable Company for between 5 and 10 years? Round your answer to the nearest tenth of a percent.

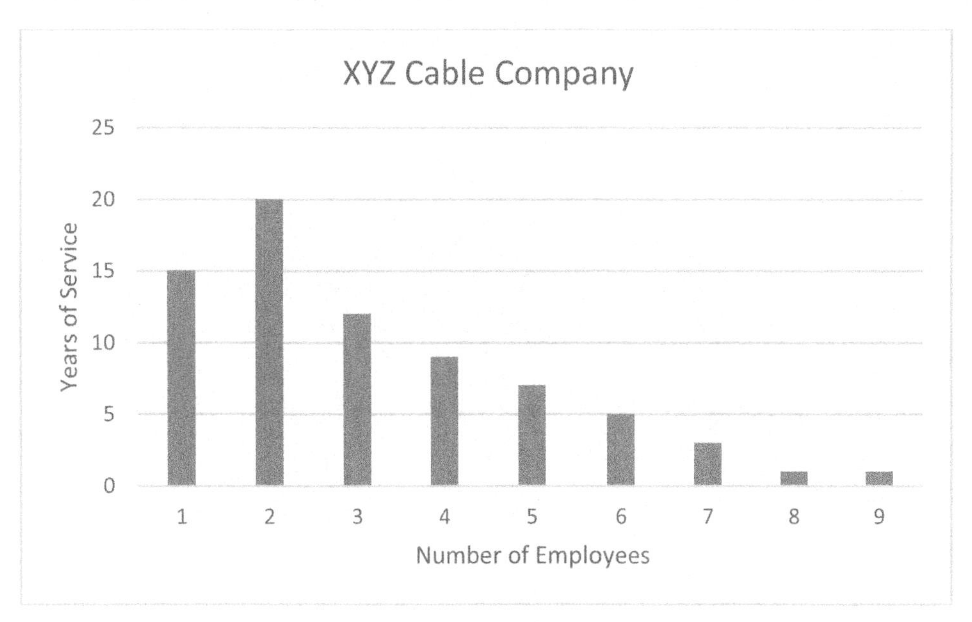

a. 11.1%
b. 33.3%
c. 84.4%
d. 8.9%

6. The chart below shows the percentage (rounded to the nearest tenth of a percent) of seniors from Heights High School and their plans for after high school. The number of seniors planning to attend a four-year college or university is 267. How many total seniors are planning to go into the military? Round your answer to the nearest integer.

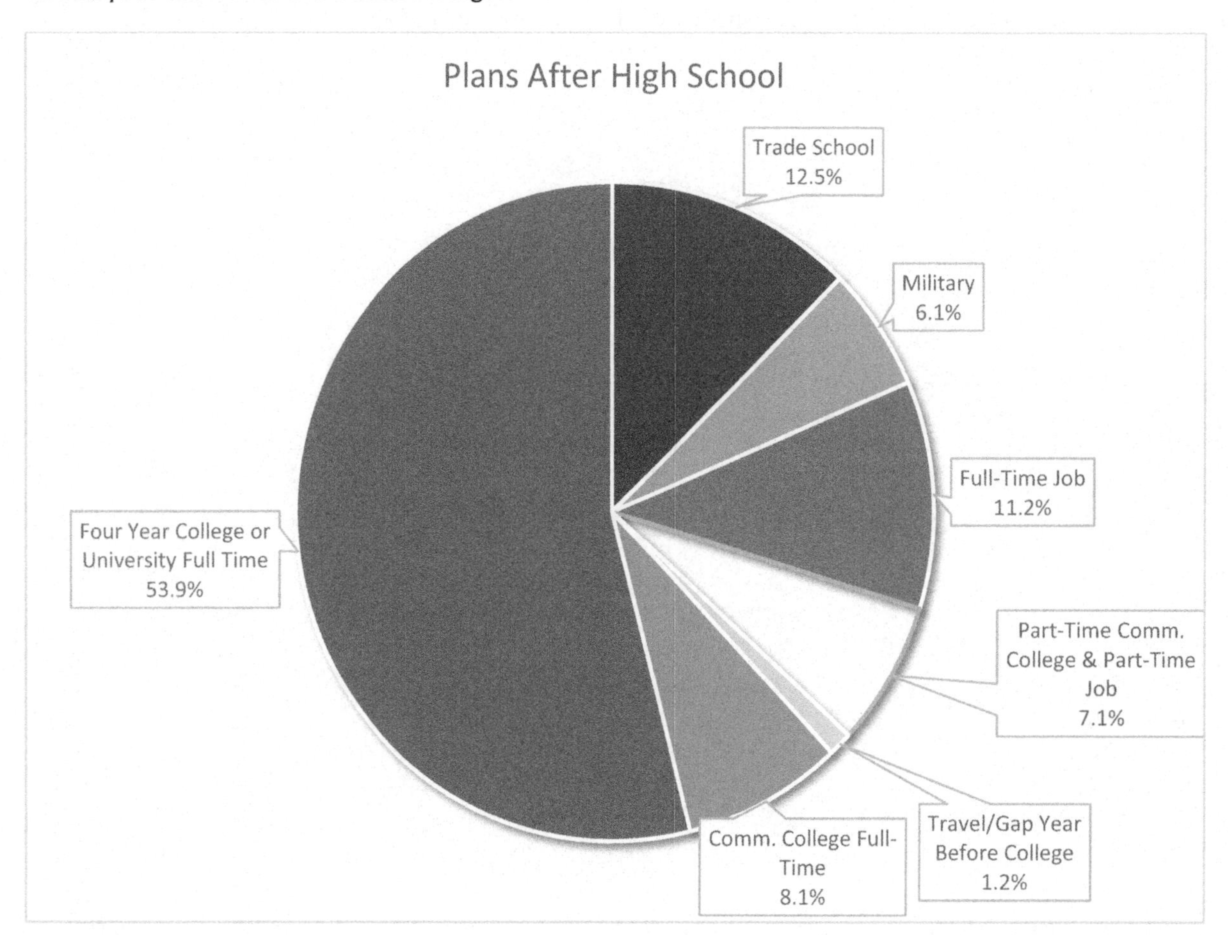

a. 495
b. 30
c. 46
d. 29

7. The graph below is a scatter plot with 30 points that correspond to the number of miles driven in a given time period for a truck driver this year. The time is given in hours. The dashed line refers to the regression line. Based on the regression line, if the truck driver drives for 7 hours, approximately how far will he go?

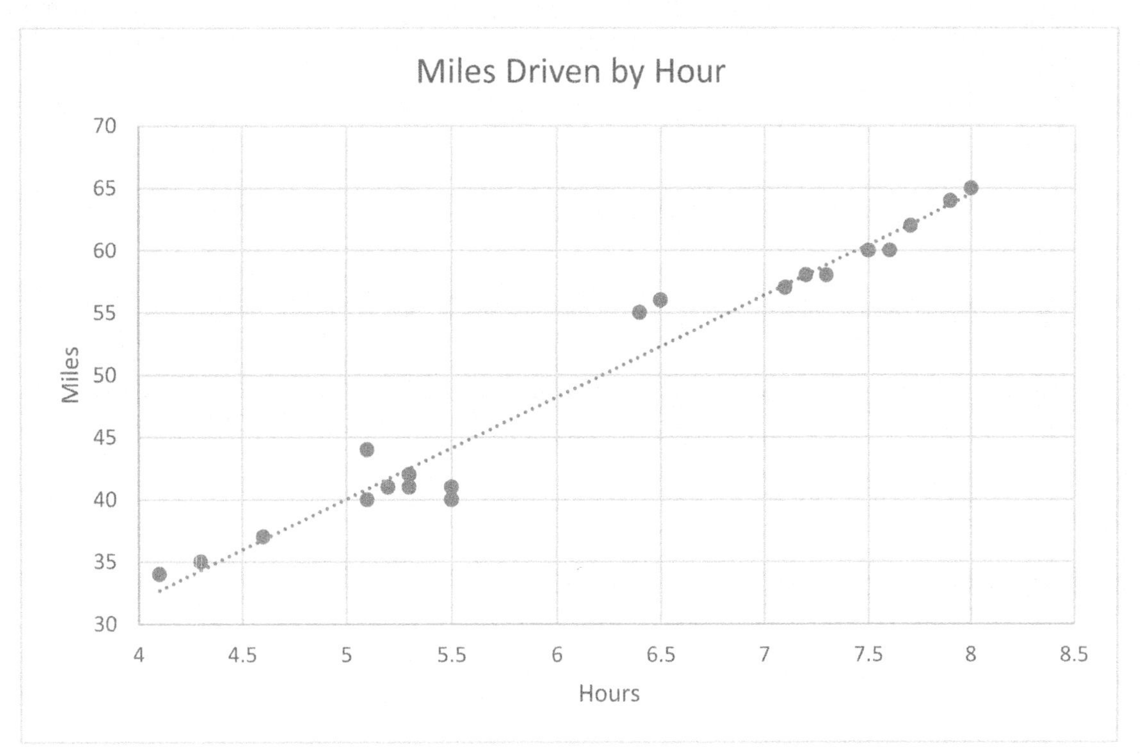

a. 55 miles
b. 57 miles
c. 60 miles
d. 50 miles

Two-Part Analysis

8. The quotient $\frac{x}{y}$ is a multiple of 9.

In the table below, choose the value of x and the value of y that are consistent with this statement. Make only one selection for x and one selection for y.

x	y	
O	O	72
O	O	119
O	O	1,296
O	O	13

9. At the state university, there are 1,350 business majors. There are 3 times as many marketing majors as accounting majors, and there are 50 students who are double majors in both accounting and marketing. The state university offers no other business degrees. In the table below, choose the number of accounting majors and the number of marketing majors consistent with this information. Make only one selection for each major.

Accounting	Marketing	
O	O	350
O	O	250
O	O	1,050
O	O	1,150

Table Analysis

10. U.S. Internet Sales in 2017 of the ABC Cable Company.

		Package Selection		
		Basic Package	Deluxe Package	Extra Deluxe Package
Region	Northeast	40.4	55.1	20.1
	Midwest	45.2	60.3	15.8
	South	34.1	41.2	23.1
	West	16.1	22.3	13.2

Note: All numbers above are in thousands.

TRUE	FALSE	
O	O	The average number of basic packages sold was 135,800.
O	O	The percentage of packages sold that were not basic was approximately 35%.
O	O	The region with the largest number of internet sales for the ABC Cable Company was the Midwest.

11.

Student	Credit Hours Completed	Major	ACT Score
Joe	18	Physical Therapy	31
Michael	16	Pre-Med	30
Kayla	14	Chemistry	28
Kendra	13	Mathematics	24
Mitchell	17	English	27
Alice	12	Nursing	31
Edward	17	Biology	32
Trista	18	Nursing	29
Kyle	19	Pre-Med	27
Stella	12	Physical Therapy	28
Rita	14	Engineering	31
Robert	14	Interior Design	30
Barry	15	Engineering	25

TRUE	FALSE	
O	O	The average number of credits completed was 16.
O	O	The probability that a student is an engineering major is higher than the probability that a student is a physical therapy major.
O	O	The data set corresponding to the ACT scores is bimodal.

12.

Year	Population of State X	Population of State Y	Population of State Z
2001	300.1	250.3	185.2
2002	301.2	251.4	185.1
2003	299.9	250.9	184.7
2004	298.8	251.1	184.9
2005	299.5	251	184.2
2006	300.2	251.7	183.9
2007	300.3	252	183.8
2008	301.1	252.3	184.5
2009	302.1	252.5	184.3
2010	301.3	252.7	185

Note: Populations shown in thousands.

TRUE	FALSE	
O	O	The largest percent change for the population in State X occurred from 2002 to 2003.
O	O	The largest percent decrease for the population in State Y occurred from 2001 to 2002.
O	O	The population in State Z increased by the same percentage from 2007 to 2008 as it decreased from 2004 to 2005.

Quantitative Reasoning

Problem Solving

1. Which of the following is the result of simplifying the expression: $\frac{4a^{-1}b^3}{a^4b^{-2}} \times \frac{3a}{b}$?
   a. $12a^3b^5$
   b. $12\frac{b^4}{a^4}$
   c. $\frac{12}{a^4}$
   d. $7\frac{b^4}{a}$
   e. $4\frac{7b}{a}$

2. What is the product of two irrational numbers?
   a. Irrational
   b. Rational
   c. Contradictory
   d. Complex and imaginary

e. Irrational or rational

3. The graph shows the position of a car over a 10-second time interval. Which of the following is the correct interpretation of the graph for the interval 1 to 3 seconds?

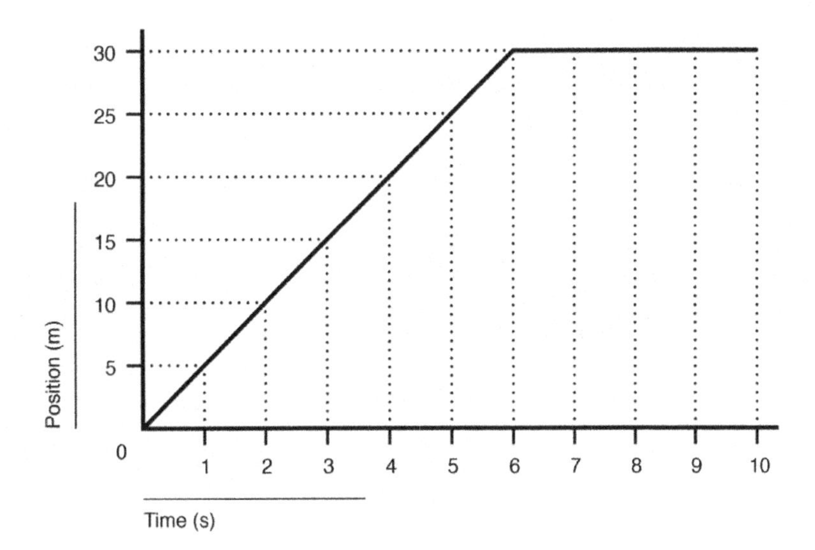

a. The car remains in the same position.
b. The car is traveling at a speed of 5m/s.
c. The car is traveling up a hill.
d. The car is traveling at 5 mph.
e. The car is traveling at a speed of 10m/s.

4. Being as specific as possible, how is the number -4 classified?
a. Real, rational, integer, whole, natural
b. Real, rational, integer, natural
c. Real, rational, integer
d. Real, irrational, complex
e. Real, irrational, whole

5. In a statistical experiment, 29 college students are given an exam during week 11 of the semester, and 30 college students are given an exam during week 12 of the semester. Both groups are being tested to determine which exam week might result in a higher grade. What's the degree of freedom in this experiment?
a. 29
b. 30
c. 59
d. 21
e. 28

6. What are the zeros of the function: $f(x) = x^3 + 4x^2 + 4x$?
   a. -2
   b. 0, -2
   c. 2
   d. 0, 2
   e. 0, 0

7. If $g(x) = x^3 - 3x^2 - 2x + 6$ and $f(x) = 2$, then what is $g(f(x))$?
   a. -26
   b. 6
   c. $2x^3 - 6x^2 - 4x + 12$
   d. -2
   e. $2x^2 + 3x + 6$

8. What is the solution to the following system of equations?

$$x^2 - 2x + y = 8$$
$$x - y = -2$$

   a. $(-2, 3)$
   b. There is no solution.
   c. $(-2, 0)$ $(1, 3)$
   d. $(-2, 0)$ $(3, 5)$
   e. $(-2, 3)$ $(3, 5)$

9. Which of the following shows the correct result of simplifying the following expression?

$$(7n + 3n^3 + 3) + (8n + 5n^3 + 2n^4)$$

   a. $9n^4 + 15n - 2$
   b. $2n^4 + 5n^3 + 15n - 2$
   c. $9n^4 + 8n^3 + 15n$
   d. $2n^4 + 8n^3 + 15n + 3$
   e. $3n^4 + 4n^3 + 15n - 4$

10. What is the product of the following expression?

$$(4x - 8)(5x^2 + x + 6)$$

   a. $20x^3 - 36x^2 + 16x - 48$
   b. $6x^3 - 41x^2 + 12x + 15$
   c. $20x^4 + 11x^2 - 37x - 12$
   d. $2x^3 - 11x^2 - 32x + 20$
   e. $10x^3 + 2x^2 - 8x + 48$

11. How could the following equation be factored to find the zeros?

$$y = x^3 - 3x^2 - 4x$$

a. $0 = x^2(x-4), x = 0, 4$
b. $0 = 3x(x+1)(x+4), x = 0, -1, -4$
c. $0 = x(x+1)(x+6), x = 0, -1, -6$
d. $0 = x^2(x-1)(x-4), x = 0, -1, 4$
e. $0 = x(x+1)(x-4), x = 0, -1, 4$

12. What is the simplified quotient of $\frac{5x^3}{3x^2y} \div \frac{25}{3y^9}$?

a. $\frac{125x}{9y^{10}}$
b. $\frac{x}{5y^8}$
c. $\frac{5}{xy^8}$
d. $\frac{xy^8}{5}$
e. $\frac{xy^2}{5}$

13. What is the solution for the following equation?

$$\frac{x^2 + x - 30}{x - 5} = 11$$

a. $x = -6$
b. There is no solution.
c. $x = 16$
d. $x = 5$
e. $x = 6$

14. Mom's car drove 72 miles in 90 minutes. How fast did she drive in feet per second?
a. 0.8 feet per second
b. 48.9 feet per second
c. 0.009 feet per second
d. 70.4 feet per second
e. 55 feet per second

15. How do you solve $V = lwh$ for h?
a. $lwV = h$
b. $h = \frac{V}{lw}$
c. $h = \frac{Vl}{w}$
d. $h = \frac{Vw}{l}$
e. $h = \frac{wl}{V}$

16. What is the domain for the function $y = \sqrt{x}$?
a. All real numbers
b. $x \geq 0$
c. $x > 0$
d. $y \geq 0$
e. $y < 0$

17. If Sarah reads at an average rate of 21 pages in four nights, how long will it take her to read 140 pages?
a. 6 nights
b. 26 nights
c. 8 nights
d. 12 nights
e. 27 nights

18. The phone bill is calculated each month using the equation $c = 50g + 75$. The cost of the phone bill per month is represented by c, and g represents the gigabytes of data used that month. What is the value and interpretation of the slope of this equation?
a. 75 dollars per day
b. 75 gigabytes per day
c. 50 dollars per day
d. 50 dollars per gigabyte
e. 25 dollars per day

19. What is the function that forms an equivalent graph to $y = \cos(x)$?
a. $y = \tan(x)$
b. $y = \csc(x)$
c. $y = \sin(x + \frac{\pi}{2})$
d. $y = \sin(x - \frac{\pi}{2})$
e. $y = \tan(x - \frac{\pi}{3})$

20. What is the solution for the equation $\tan(x) + 1 = 0$, where $0 \leq x < 2\pi$?

a. $x = \frac{3\pi}{4}, \frac{5\pi}{4}$

b. $x = \frac{3\pi}{4}, \frac{\pi}{4}$

c. $x = \frac{5\pi}{4}, \frac{7\pi}{4}$

d. $x = \frac{3\pi}{4}, \frac{7\pi}{4}$

e. $x = \frac{3\pi}{2}, \frac{7\pi}{2}$

Data Sufficiency

Decide whether the data given in the statements are sufficient to answer the question.

21. Katie is delivering 4 packages today. In how many ways can she select the packages that she delivers?

1) She never delivers more than 5 packages at once.
2) Her employer has 10 packages available to deliver.

a. Statement (1) ALONE is sufficient, but statement (2) ALONE is NOT sufficient.
b. Statement (2) ALONE is sufficient, but statement (1) ALONE is NOT sufficient.
c. BOTH statements TOGETHER are sufficient, but NEITHER statement ALONE is sufficient.
d. EACH statement ALONE is sufficient.
e. Statements (1) and (2) TOGETHER are NOT sufficient.

22. Katie and Serena are writing a story for a newspaper. Katie can write at a speed 2 times faster than Serena can. How long will it take them to finish the story while working together?

1) Working individually, Katie could finish the story in 4 hours.
2) Serena can write 1,500 words per hour.

a. Statement (1) ALONE is sufficient, but statement (2) ALONE is NOT sufficient.
b. Statement (2) ALONE is sufficient, but statement (1) ALONE is NOT sufficient.
c. BOTH statements TOGETHER are sufficient, but NEITHER statement ALONE is sufficient.
d. EACH statement ALONE is sufficient.
e. Statements (1) and (2) TOGETHER are NOT sufficient.

23. Is y a positive number?

1) $y^2 > 25$
2) $y + 6 > 10$

a. Statement (1) ALONE is sufficient, but statement (2) ALONE is NOT sufficient.
b. Statement (2) ALONE is sufficient, but statement (1) ALONE is NOT sufficient.
c. BOTH statements TOGETHER are sufficient, but NEITHER statement ALONE is sufficient.
d. EACH statement ALONE is sufficient.
e. Statements (1) and (2) TOGETHER are NOT sufficient.

24. Does y = z?

1) $\frac{y}{z} < 1$
2) $\frac{z-y}{2} < -1$

a. Statement (1) ALONE is sufficient, but statement (2) ALONE is NOT sufficient.
b. Statement (2) ALONE is sufficient, but statement (1) ALONE is NOT sufficient.
c. BOTH statements TOGETHER are sufficient, but NEITHER statement ALONE is sufficient.
d. EACH statement ALONE is sufficient.
e. Statements (1) and (2) TOGETHER are NOT sufficient.

25. If $a + b + c = 160$, what is a?

1) $b = 80$
2) $b + c = 120$

a. Statement (1) ALONE is sufficient, but statement (2) ALONE is NOT sufficient.
b. Statement (2) ALONE is sufficient, but statement (1) ALONE is NOT sufficient.
c. BOTH statements TOGETHER are sufficient, but NEITHER statement ALONE is sufficient.
d. EACH statement ALONE is sufficient.
e. Statements (1) and (2) TOGETHER are NOT sufficient.

26. Is $9 - \frac{x}{9}$ an integer?

1) x is a multiple of 6.
2) The process of the division of x by 9 results in a zero remainder.

a. Statement (1) ALONE is sufficient, but statement (2) ALONE is NOT sufficient.
b. Statement (2) ALONE is sufficient, but statement (1) ALONE is NOT sufficient.
c. BOTH statements TOGETHER are sufficient, but NEITHER statement ALONE is sufficient.
d. EACH statement ALONE is sufficient.
e. Statements (1) and (2) TOGETHER are NOT sufficient.

27. To stock your grocery store, you purchase boxes of oatmeal from your supplier for $2 each and sell them to customers. How much do you individually sell them for?

1) Your profit from each sale is 25%.
2) The amount that you pay for the oatmeal is 80% of what you sell it for.

a. Statement (1) ALONE is sufficient, but statement (2) ALONE is NOT sufficient.
b. Statement (2) ALONE is sufficient, but statement (1) ALONE is NOT sufficient.
c. BOTH statements TOGETHER are sufficient, but NEITHER statement ALONE is sufficient.
d. EACH statement ALONE is sufficient.
e. Statements (1) and (2) TOGETHER are NOT sufficient.

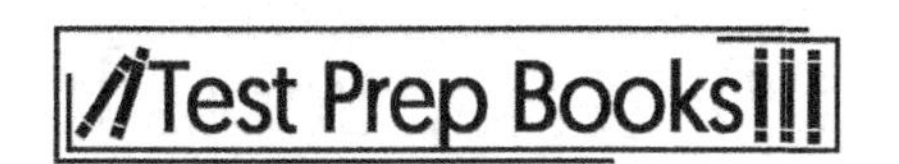

28. Ten square tiles are to be placed together to tile a kitchen pantry. What is the perimeter of the pantry?

1) The area of each tile is 16 square inches.
2) The area of the pantry is 160 square inches.

a. Statement (1) ALONE is sufficient, but statement (2) ALONE is NOT sufficient.
b. Statement (2) ALONE is sufficient, but statement (1) ALONE is NOT sufficient.
c. BOTH statements TOGETHER are sufficient, but NEITHER statement ALONE is sufficient.
d. EACH statement ALONE is sufficient.
e. Statements (1) and (2) TOGETHER are NOT sufficient.

29. If $x \neq 0$, is $x^4 < x$?

1) $x < 1$
2) $x < -2$

a. Statement (1) ALONE is sufficient, but statement (2) ALONE is NOT sufficient.
b. Statement (2) ALONE is sufficient, but statement (1) ALONE is NOT sufficient.
c. BOTH statements TOGETHER are sufficient, but NEITHER statement ALONE is sufficient.
d. EACH statement ALONE is sufficient.
e. Statements (1) and (2) TOGETHER are NOT sufficient.

30. A set consists of the total snowfall for 7 days. Is the median of the set less than the mode?

1) The mean and median of the set are equal to 3.
2) The mode appears twice in the set.

a. Statement (1) ALONE is sufficient, but statement (2) ALONE is NOT sufficient.
b. Statement (2) ALONE is sufficient, but statement (1) ALONE is NOT sufficient.
c. BOTH statements TOGETHER are sufficient, but NEITHER statement ALONE is sufficient.
d. EACH statement ALONE is sufficient.
e. Statements (1) and (2) TOGETHER are NOT sufficient.

31. The dimensions of a rectangular box are 7 centimeters by 11 centimeters by 12 centimeters. If a cylinder is placed inside the box so that either circular end touches the top and bottom of the box, what is the volume of the circular cylinder?

1) The surface area of the cylinder is 128π square centimeters.
2) The base of a box has an area of 77 square centimeters.

a. Statement (1) ALONE is sufficient, but statement (2) ALONE is NOT sufficient.
b. Statement (2) ALONE is sufficient, but statement (1) ALONE is NOT sufficient.
c. BOTH statements TOGETHER are sufficient, but NEITHER statement ALONE is sufficient.
d. EACH statement ALONE is sufficient.
e. Statements (1) and (2) TOGETHER are NOT sufficient.

32. In the Cartesian coordinate system, does the point (-2, 10) lie on a given line?

1) The point (4, -2) is a point on the given line.
2) The point (1, 4) is a point on the given line.

a. Statement (1) ALONE is sufficient, but statement (2) ALONE is NOT sufficient.
b. Statement (2) ALONE is sufficient, but statement (1) ALONE is NOT sufficient.
c. BOTH statements TOGETHER are sufficient, but NEITHER statement ALONE is sufficient.
d. EACH statement ALONE is sufficient.
e. Statements (1) and (2) TOGETHER are NOT sufficient.

33. Lines l and m intersect at one point. Are the 2 lines perpendicular?

1) The slope of line l is 5.
2) The equation of line m is $x+5y=2$.

a. Statement (1) ALONE is sufficient, but statement (2) ALONE is NOT sufficient.
b. Statement (2) ALONE is sufficient, but statement (1) ALONE is NOT sufficient.
c. BOTH statements TOGETHER are sufficient, but NEITHER statement ALONE is sufficient.
d. EACH statement ALONE is sufficient.
e. Statements (1) and (2) TOGETHER are NOT sufficient.

34. A bag contains only red, green, and blue marbles. In total, there are 39 marbles in the bag. What is the probability of pulling a red marble from the bag?

1) There are 14 blue marbles.
2) The probability of selecting a green marble is equal to the probability of selecting a blue marble.

a. Statement (1) ALONE is sufficient, but statement (2) ALONE is NOT sufficient.
b. Statement (2) ALONE is sufficient, but statement (1) ALONE is NOT sufficient.
c. BOTH statements TOGETHER are sufficient, but NEITHER statement ALONE is sufficient.
d. EACH statement ALONE is sufficient.
e. Statements (1) and (2) TOGETHER are NOT sufficient.

35. In a data set that is normally distributed, what is the standard deviation?

1) The mean of the data set is 37.1 and the data point that is one standard deviation below the mean is 33.4.
2) The data point that is one standard deviation above the mean is 40.8 and the data point that is 2 standard deviations above the mean is 44.5.

a. Statement (1) ALONE is sufficient, but statement (2) ALONE is NOT sufficient.
b. Statement (2) ALONE is sufficient, but statement (1) ALONE is NOT sufficient.
c. BOTH statements TOGETHER are sufficient, but NEITHER statement ALONE is sufficient.
d. EACH statement ALONE is sufficient.
e. Statements (1) and (2) TOGETHER are NOT sufficient.

36. If $a \neq b, d \neq c$, and $b \neq d$, is

$$\frac{a^6(a-b)^4(c-d)(d-b)}{(d-c)(a-b)^2(d-b)^4} > 0?$$

1) $b > d$
2) $a > 0$

a. Statement (1) ALONE is sufficient, but statement (2) ALONE is NOT sufficient.
b. Statement (2) ALONE is sufficient, but statement (1) ALONE is NOT sufficient.
c. BOTH statements TOGETHER are sufficient, but NEITHER statement ALONE is sufficient.
d. EACH statement ALONE is sufficient.
e. Statements (1) and (2) TOGETHER are NOT sufficient.

37. If $ab \neq 0$, is $a > b$?

1) $|a| > |b|$
2) $a = 4b$.

a. Statement (1) ALONE is sufficient, but statement (2) ALONE is NOT sufficient.
b. Statement (2) ALONE is sufficient, but statement (1) ALONE is NOT sufficient.
c. BOTH statements TOGETHER are sufficient, but NEITHER statement ALONE is sufficient.
d. EACH statement ALONE is sufficient.
e. Statements (1) and (2) TOGETHER are NOT sufficient.

Verbal Reasoning

Reading Comprehension

Questions 1–4 are based on the following passage:

An Organization for Economic Cooperation and Development study of ten developing countries during the period from 1985 to 1992 found significant implementation of privatization in only three countries. The study concluded that "reductions in the central budget deficit can only be marginal" because the impact was not evaluated over several years to consider the effect of the revenues foregone from state-owned enterprises (SOEs). Several later studies measured the budgetary effects and reported significant increases in profitability and productivity as a result of privatization, but the methodological flaws related to the difficulty of isolating the performance of SOEs from other elements rendered the findings ambiguous. While the evidence on the performance of SOEs "shows that state ownership is often correlated with politicization, inefficiency, and waste of resources," the assumption that it is state ownership that creates an environment influencing the quality of performance is not proven, with the empirical research on this point having yielded conflicting results. Given the inconclusive evidence, many scholars did not concur on a World Bank statement in 1995 that SOEs "remain an important obstacle to better economic performance."

Reflecting a belief that the market is the best allocator of resources, experts have often recommended "unleashing" the private sector by removing regulations and privatizing SOEs. In

1995, to preclude hasty and simplistic privatization efforts, the World Bank recommended that SOEs be corporatized under commercial law and issued guidance on "[p]re-privatization interim measures and institutional arrangements for 'permanent SOEs.'" The bank also listed five preconditions for successful privatization: hard budget constraints; capital and labor market discipline; competition; corporate governance free of political interference; and commitment to privatization.

In view of the pervasive presence of SOEs in the global economy and their embodiment of political and economic considerations, SOEs are an entity to be considered and managed in the pursuit of stability.

"The State-Owned Enterprise as a Vehicle for Stability" by Neil Efird (2010), published by the Strategic Studies Institute (Department of Defense), pgs. 7–8

1. The World Bank issued which statement(s) in 1995?
   a. State-owned enterprises cannot impede economic performance.
   b. State-owned enterprises play a critical role in developing countries.
   c. State-owned enterprises promote economic and political stability.
   d. State-owned enterprises should be corporatized under commercial law gradually.
   e. State-owned enterprises might be inefficient, but the evidence is inconclusive.

2. Based on the passage above, which statement(s) can be properly inferred?
   a. State-owned enterprises always cause economic stagnation.
   b. Privatization is controversial, even among economic experts.
   c. Economic studies are always subject to intense criticism and secondhand guessing.
   d. State-owned enterprises violate commercial law.
   e. The World Bank holds the power to directly intervene in economies.

3. Which statement(s) about state-owned enterprises is true based on the passage above?
   a. The empirical research demonstrates that state-owned enterprises are efficient and productive.
   b. Developing countries have little influence on the World Bank's policies.
   c. Privatization enjoys widespread popular support wherever it is implemented.
   d. State-owned enterprises do not have sizable effects on the global economy.
   e. Among other factors, successful privatization requires competition and corporate governance that is free of political interference.

4. Which statement(s) most accurately identifies the author's ultimate conclusion?
   a. The market is the best allocator of resources, so private enterprises will always outperform state-owned enterprises.
   b. The World Bank holds considerable expertise in matters related to state-owned enterprises and privatization.
   c. State-owned enterprises should be managed in a way that promotes economic stability, which might require a measured approach to privatization.
   d. Studies conducted with a flawed methodology should not be the basis for economic decisions.
   e. State-owned enterprises should be privatized under commercial law as long as the government adheres to the five preconditions for privatization.

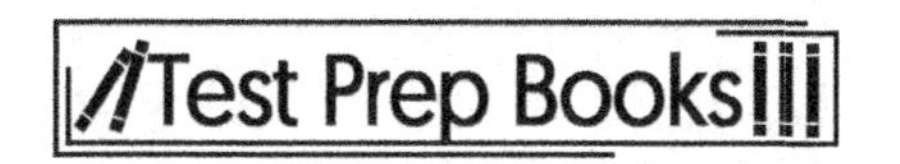

Questions 5–8 are based on the following passage:

> Scholars who have examined how national leaders historically craft public speeches in response to accusations of offensive words or deeds conclude that such officials generally rely upon one of two recurrent strategic approaches. The first of these is apologia, which William Benoit and Susan Brinson define as "a recurring type of discourse designed to restore face, image, or reputation after an alleged or suspected wrongdoing," which occurs during many apologies. The second is reconciliation, which John Hatch defines as "a dialogic rhetorical process of healing between the parties."
>
> Speakers using apologia strive to restore their own credibility and remove any perception they might be guilty of involvement in the transgression. Speakers seeking reconciliation are interested in restoring dialogue instead of pursuing the purposes of shifting blame, denying charges, or some other form of blame avoidance or image repair.
>
> Apologia focuses on short-term gains achievable by regaining favor with audiences already predisposed to the speaker's arguments. Reconciliation, by contrast, has a goal of understanding the long-term processes of image restoration and mutual respect between the aggrieved and the transgressor. Attempts at credible reconciliation utilize symbols of reunion to demonstrate that the aggrieved has *genuinely* granted the forgiveness sought by the offender. Visual images freeze the moment of genuine forgiveness and, when replayed in the online environment, carry forward the steps of reunification into perpetuity.
>
> Not all offending images circulating in the online environment warrant a visual response or even an apology by national leaders. Nevertheless, failure to respond to the small number of potent images of transgressions that share characteristics qualifying them for continued recirculation in future propaganda efforts could be a costly mistake. Reconciliation provides a fruitful choice, as its long-term goals match the ongoing need to handle ever-circulating images that offend.

Visual Propaganda and Extremism the Online Environment, edited by Carol K. Winkler and Cori E. Dauber (2014), published by the Strategic Studies Institute (Department of Defense), "Visual Reconciliation as Strategy of Response to Offending Images" by Carol K. Winkler, excerpted from pages 63-64, 71-72, and 74

5. Which statement(s) accurately describes the difference between apologia and reconciliation based on the passage?

a. Apologia is a dialogic rhetorical process that involves healing, while reconciliation seeks to restore the image of a party suspected of wrongdoing.
b. Reconciliation utilizes symbols to demonstrate forgiveness, while apologia is always delivered in writing.
c. Apologia is more effective in the short term, while reconciliation is more useful for improving relationships in the long term.
d. Reconciliation seeks to close channels of communication, while apologia is more closely related to shifting blame and rehabilitating credibility.
e. Apologia requires the aggrieved party's forgiveness, while reconciliation is a strategic approach to delivering visual responses.

6. Based on the passage above, which statement(s) can be properly inferred?
   a. National leaders should never apologize because reconciliation is always the superior option.
   b. Reconciliation is more effective than apologia because it is not self-serving.
   c. Apologia and reconciliation are most effective when delivered together.
   d. Apologia and reconciliation only function properly in the online environment.
   e. Maintaining a positive national image is an important part of governance.

7. Which statement(s) describes a feature of reconciliation?
   a. Reconciliation involves restoring dialogue and repairing relationships in the short term.
   b. Reconciliation leverages symbols of reunion to demonstrate that the aggrieved has genuinely granted the forgiveness sought by the offender.
   c. Reconciliation focuses on shifting blame, denying charges, restoring credibility, or otherwise repairing the offender's image.
   d. Reconciliation should be used when the audience is already predisposed to the speaker's argument.
   e. Reconciliation prioritizes short-term gains over substantively altering the underlying relationship.

8. Which statement(s) describes how the author believes national leaders should respond to offending images circulating in the online environment?
   a. National leaders should censor all offending images and attack any group that disseminates propaganda.
   b. National leaders should conduct an anti-propaganda campaign to raise public awareness about the dangers of sensationalism.
   c. The circulation of offending images in online environments is inevitable but relatively harmless, so this phenomenon should be ignored.
   d. All offending images should be addressed with either apologia or reconciliation, depending on which is more appropriate based on context.
   e. Some offending images do not warrant any response, but for the ones that do, reconciliation is the more appealing option due to its long-term impact.

Questions 9–11 are based on the following passage:

> Despite spending far more on health care than any other nation, the United States ranks near the bottom on key health indicators. This paradox has been attributed to underinvestment in addressing social and behavioral determinants of health. A recent Institute of Medicine (IOM) report linked the shorter overall life expectancy in the United States to problems that are either caused by behavioral risks (e.g., injuries and homicides, adolescent pregnancy and sexually transmitted infections (STIs), HIV/AIDS, drug-related deaths, lung diseases, obesity, and diabetes) or affected by social conditions (e.g., birth outcomes, heart disease, and disability).
>
> While spending more than other countries per capita on health care services, the United States spends less on average than do other nations on social services impacting social and behavioral determinants of health. Bradley, et al., found that Organization for Economic Co-operation and Development (OECD) nations with a higher ratio of spending on social services relative to health care services have better health and longer life expectancies than do those like the United States that have a lower ratio.
>
> The Clinical & Translational Science Awards (CTSAs) established by the National Institutes of Health (NIH) have helped initiate interdisciplinary programs in more than sixty institutions that

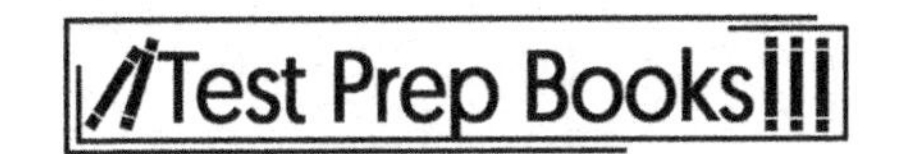

aim to advance the translation of research findings from "bench" to "bedside" to "community." Social and behavioral issues are inherent aspects of the translation of findings at the bench into better care and better health. Insofar as Clinical and Translational Science Institutes (CTSIs) will be evaluated for renewal—not only on the basis of their bench science discoveries, but also by their ability to move these discoveries into practice and improve individual and population health—the CTSIs should be motivated to include social and behavioral scientists in their work.

Population Health: Behavioral and Social Science Insights, Robert M. Kaplan et al. (2015), published by the Agency for Healthcare Research and Quality (National Institutes of Health), "Determinants of Health and Longevity" by Nancy E. Adler and Aric A. Prather, excerpted from pages 411 and 417

9. Which statement(s) describes how the United States differs in its approach to health care compared with other nations?

a. On average, the United States spends more on social services impacting social and behavioral determinants of health than other nations.
b. Compared with other nations, the United States has a higher ratio of spending on social services relative to health care services.
c. Compared with other nations, the United States spends more per capita on health care services despite producing worse health outcomes.
d. Compared with other nations, the United States has a lower life expectancy due to its lack of spending on health care services.
e. Unlike other nations, the United States doesn't fund interdisciplinary programs that include behavioral and social science.

10. Based on the passage, which statement(s) describes the primary purpose of the Clinical and Translational Science Awards?

a. The Clinical and Translational Science Awards advocate for the expansion of social services in the United States.
b. The Clinical and Translational Science Awards seek to advance the translation of research findings from "bench" to "bedside" to "community."
c. The Clinical and Translational Science Awards conduct research on how to mitigate the behavioral risks that have caused the decline in Americans' life expectancy.
d. The Clinical and Translational Science Awards exclusively employ social and behavioral scientists, filling a void in the American health care system.
e. The Clinical and Translational Science Awards calculate the optimal ratio for spending on social services relative to health care services.

11. Which statement(s) most accurately identifies the author's main thesis?

a. Despite outspending other countries on health care, the United States performs poorly on key health indicators.
b. The National Institutes of Health created the Clinical and Translational Science Awards to develop interdisciplinary programs in more than sixty institutions.
c. Social and behavioral factors are an underappreciated aspect of health, and if they are better understood and properly addressed, health outcomes will improve.
d. The United States has the shortest overall life expectancy in the world due to unaddressed behavioral risks and deteriorating social conditions.
e. Life expectancy is the most important health care indicator because it encapsulates every other relevant factor.

Questions 12–14 are based on the following passages.

Passage I

Lethal force, or deadly force, is defined as the physical means to cause death or serious harm to another individual. The law holds that lethal force is only accepted when you or another person are in immediate and unavoidable danger of death or severe bodily harm. For example, a person could be beating a weaker person in such a way that they are suffering severe enough trauma that could result in death or serious harm. This would be an instance where lethal force would be acceptable and possibly the only way to save that person from irrevocable damage.

Another example of when to use lethal force would be when someone enters your home with a deadly weapon. The intruder's presence and possession of the weapon indicate mal-intent and the ability to inflict death or severe injury to you and your loved ones. Again, lethal force can be used in this situation. Lethal force can also be applied to prevent the harm of another individual. If a woman is being brutally assaulted and is unable to fend off an attacker, lethal force can be used to defend her as a last-ditch effort. If she is in immediate jeopardy of rape, harm, and/or death, lethal force could be the only response that could effectively deter the assailant.

The key to understanding the concept of lethal force is the term *last resort*. Deadly force cannot be taken back; it should be used only to prevent severe harm or death. The law does distinguish whether the means of one's self-defense is fully warranted, or if the individual goes out of control in the process. If you continually attack the assailant after they are rendered incapacitated, this would be causing unnecessary harm, and the law can bring charges against you. Likewise, if you kill an attacker unnecessarily after defending yourself, you can be charged with murder. This would move lethal force beyond necessary defense, making it no longer a last resort but rather a use of excessive force.

Passage II

Assault is the unlawful attempt of one person to apply apprehension on another individual by an imminent threat or by initiating offensive contact. Assaults can vary, encompassing physical strikes, threatening body language, and even provocative language. In the case of the latter, even if a hand has not been laid, it is still considered an assault because of its threatening nature.

Let's look at an example: A homeowner is angered because his neighbor blows fallen leaves into his freshly mowed lawn. Irate, the homeowner gestures a fist to his fellow neighbor and threatens to bash his head in for littering on his lawn. The homeowner's physical motions and verbal threat heralds a physical threat against the other neighbor. These factors classify the homeowner's reaction as an assault. If the angry neighbor hits the threatening homeowner in retaliation, that would constitute an assault as well because he physically hit the homeowner.

Assault also centers on the involvement of weapons in a conflict. If someone fires a gun at another person, it could be interpreted as an assault unless the shooter acted in self-defense. If an individual drew a gun or a knife on someone with the intent to harm them, it would be considered assault. However, it's also considered an assault if someone simply aimed a weapon, loaded or not, at another person in a threatening manner.

12. The purpose of the second passage is to:
   a. To inform the reader about what assault is and how it is committed
   b. To inform the reader about how assault is a minor example of lethal force
   c. To disprove the previous passage concerning lethal force
   d. To argue that the use of assault is more common than the use of lethal force
   e. To recount an incident in which the author was assaulted

13. Given the information in the passages, one true thing about assault is that:
   a. All assault is considered an expression of lethal force.
   b. There are various forms of assault.
   c. Smaller, weaker people cannot commit assault.
   d. Assault is justified only as a last resort.
   e. Assault charges are more severe than unnecessary use of force charges.

14. The way in which the passages are structured is that:
   a. Both passages open by defining a legal concept and then continue to describe situations in order to further explain the concept.
   b. Both passages begin with situations, introduce accepted definitions, and then cite legal ramifications.
   c. The first passage presents a long definition while the second passage begins by showing an example of assault.
   d. Both cite specific legal doctrines, then proceed to explain the rulings.
   e. The first passage explains both concepts and then focuses on lethal force. The second passage picks up with assault and explains the concept in depth.

Critical Reasoning

15. Teacher: Students don't need parental involvement to succeed. In my class of twenty kids, the two highest achieving students come from foster homes. There are too many children in the foster homes for their parents to monitor homework and enforce study habits. It's always the case that students can overcome their parents' indifference.

What mistake does the teacher commit in his reasoning?
   a. The teacher incorrectly applies a common rule.
   b. The teacher's conclusion is totally unjustified.
   c. The teacher relies on an unreasonably small sample size in drawing his conclusion.
   d. The teacher fails to consider competing theories.
   e. The teacher is biased.

16. Trent is a member of the SWAT Team, the most elite tactical unit at the city's police department. SWAT apprehends more suspected criminals than all other police units combined. Taken as a whole, the police department solves a higher percentage of crime than ever before in its history. Within the SWAT team, Trent's four-man unit is the most successful. However, the number of unsolved crime increases every year.

Which of the following statements, if true, most logically resolves the apparent paradox?

a. Trent's SWAT team is the city's best police unit.
b. Violent crime has decreased dramatically, while petty drug offenses have increased substantially.
c. The total number of crimes increases every year.
d. Aside from the SWAT units, the police department is largely incompetent.
e. The police department focuses more on crimes involving serious injury or significant property damage.

17. Scientist: The FDA has yet to weigh in on the effects of electronic cigarettes on long-term health. Electronic cigarettes heat up a liquid and produce the vapor inhaled by the user. The liquid consists of vegetable glycerin and propylene glycerol in varying ratios. Artificial flavoring is also added to the liquid. Although the FDA has approved vegetable glycerin, propylene glycerol, and artificial flavors for consumption, little is known about the effects of consuming their vapors. However, electronic cigarettes do not produce tar, which is one of the most dangerous chemicals in tobacco cigarettes.

Which one of the following most accurately expresses the scientist's main point?

a. The FDA is inefficient and ineffective at protecting public health.
b. Electronic cigarettes' liquid is probably safer than tobacco.
c. Smokers should quit tobacco and start using electronic cigarettes.
d. Tar is the reason why cigarettes are unhealthy.
e. Although all of the information is not yet available, electronic cigarettes are promising alternatives to tobacco since the former does not produce tar.

18. All Labrador retrievers love playing fetch. Only German shepherds love protecting their home. Some dogs are easy to train. Brittany's dog loves playing fetch and loves protecting her home.

Which one of the following statements must be true?

a. Brittany's dog is a Labrador retriever.
b. Brittany's dog is a German shepherd.
c. Brittany's dog is easy to train.
d. Brittany's dog is half Labrador retriever and half German shepherd.
e. Brittany's dog is half Labrador retriever and half German shepherd, and her dog is also easy to train.

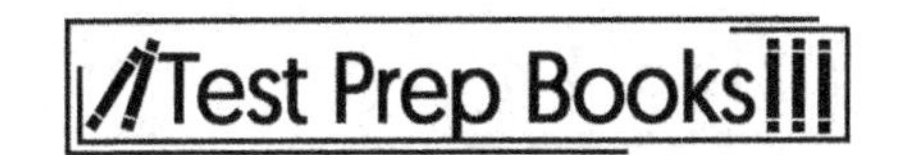

19. Sociologist: Poverty is the number one cause of crime. When basic needs, like food and shelter, are not met, people become more willing to engage in criminal activity. The easiest way to reduce crime is to lessen poverty.

Which one of the following statements, if true, best supports the sociologist's argument?

a. The typical criminal is less wealthy than the average person.
b. The easiest way to lessen poverty is to redistribute wealth.
c. Drug addiction and substance abuse is the second largest cause of crime, and drug users who struggle with addiction are more impoverished than the average person.
d. Moral societies should guarantee that all their members' basic needs are met.
e. Studies show that most crimes involve food theft and trespassing.

20. Regular weightlifting is necessary for good health. Weightlifting with heavy resistance, especially with compound movements, helps break down and rebuild stronger muscle fibers, resulting in strength and size gains.

Which one of the following is an assumption required by the argument?

a. Strength and size gains are indicators of good health.
b. Compound movements are the only way to increase strength and size.
c. Performing compound movements is necessary for good health.
d. Performing compound movements is the only way to break down and rebuild stronger muscle fibers.
e. Regular weightlifting is necessary for good health.

21. West Korea's economy is experiencing high rates of growth for the sixth consecutive quarter. An autocratic despot dominates all aspects of West Korean society, and as a result, West Koreans enjoy less civil liberties and freedom than neighboring countries. Clearly, civil liberties do not impact economic gains.

The following, if true, strengthens the argument, EXCEPT:

a. Neighboring countries' democratic processes are often deadlocked and unable to respond to immediate economic problems.
b. The autocratic despot started governing the country six quarters ago.
c. West Korea found a massive oil reserve under the country shortly before the autocratic despot seized power.
d. Political protests in neighboring countries often shorten workdays and limit productivity.
e. The West Korean autocratic despot devotes all of his time to solving economic problems.

22. Sociologist: Marriage is one of the most important societal institutions. The marital relationship provides numerous structural benefits for married couples and their offspring. Studies consistently show that children born out of wedlock are less likely to attend college and more likely to work low-paying jobs. Additionally, married people are more likely to be homeowners and save for retirement. Therefore, if marriage rates decline, _______________.

Which one of the following most logically completes the sociologist's argument?

a. Society will collapse.
b. Everyone would have less money.
c. Nobody would own homes.
d. People would be happier.
e. College attendance would probably decline.

23. Economist: Countries with lower tax rates tend to have stronger economies. Although higher taxes raise more revenue, highly taxed consumers have less disposable income. An economy can never grow if consumers aren't able to purchase goods and services. Therefore, the government should lower tax rates across the board.

The economist's argument depends on assuming that:

a. The top five world economies have the lowest tax rates in the world.
b. Consumers' disposable income is directly related to their ability to purchase goods and services.
c. Lower tax rates will be much more popular with consumers.
d. Increasing disposable income is the only way to ensure economic growth.
e. Economic growth is more important than supporting social welfare programs.

24. The United States' economy continues to grow. Over the last decade, the country's Gross Domestic Product—the monetary value of all finished goods and services produced within a country's borders—has increased by between 2 and 4 percent. The United States' economy is guaranteed to grow between 2 and 4 percent next year.

The flawed reasoning in which of the following arguments most mirrors the flawed reasoning presented in the argument above?

a. Ted is obsessed with apple pie. He's consumed one whole pie every day for the last decade. Ted will probably eat a whole apple pie tomorrow.
b. Last year Alexandra finished as the top salesperson at her company. She will undoubtedly be the top salesperson next year.
c. George always brushes his teeth right before getting into bed. His bedtime routine has remained the same for two decades. It's more probable than not that George brushes his teeth right before getting into bed tomorrow night.
d. Germany's economy is the strongest it's been since the end of World War II. Over the last decade, the country's Gross Domestic Product—the monetary value of all finished goods and services produced within a country's borders—has increased by between 2 and 4 percent. Germany's economic growth is a result of inclusive democratic processes.
e. Tito is the top ranked surfer in the world. Las Vegas bookmakers listed him as a clear favorite to win the upcoming invitational tournament. Tito is more likely to win the invitational than any other surfer.

25. Zookeeper: Big cats are undoubtedly among the smartest land mammals. Lions, tigers, and jaguars immediately adjust to their new surroundings. Other animals refuse to eat or drink in captivity, but the big cats relish their timely prepared meals. Big cats never attempt to escape their enclosures.

Which one of the following, if true, most weakens the zookeeper's argument?

a. Big cats don't attempt to escape because they can't figure out their enclosures' weak spots.
b. No qualified expert believes that adjusting to captivity is a measure of intelligence.
c. Bears also do not have any trouble adjusting to captivity.
d. A recent study comparing the brain scans of large mammals revealed that big cats exhibit the most brain activity when stimulated.
e. Zoos devote exponentially more resources to big cats relative to other animals.

26. Tanya is a lawyer. Nearly all lawyers dutifully represent their clients' best interests, but a few unethical ones charge exorbitant and fraudulent fees for services. Some lawyers become millionaires, while others work in the best interest of the public. However, all lawyers are bound by extensive ethical codes, which vary slightly by jurisdiction.

If the statements above are true, which one of the following must also be true?

a. Tanya dutifully represents her clients' best interests.
b. Tanya charges exorbitant fees for her services.
c. Tanya is bound by extensive ethical codes.
d. Tanya is a millionaire.
e. Tanya works for the public sector.

Sentence Correction

27. Early in my career, a master's teacher shared this thought with me "Education is the last bastion of civility."

a. a master's teacher shared this thought with me "Education is the last bastion of civility."
b. a master's teacher shared this thought with me: "Education is the last bastion of civility."
c. a master's teacher shared this thought with me: "Education is the last bastion of civility".
d. a master's teacher shared this thought with me. "Education is the last bastion of civility."
e. a master' teacher shared this thought with me; "Education is the last bastion of civility."

28. Which of the following words is spelled incorrectly?

It is really what makes us human and what distinguishes us as civilised creatures.

a. makes
b. human
c. distinguishes
d. civilised
e. creatures

29. Education should never discriminate on any basis, and it should create individuals who are self-sufficient, patriotic, and tolerant of others' ideas.

a. others' ideas
b. other's ideas
c. others ideas
d. others's ideas
e. "others" ideas

30. All children can learn. Although not all children learn in the same manner.

a. All children can learn. Although not all children learn in the same manner.
b. All children can learn although not all children learn in the same manner.
c. All children can learn although, not all children learn in the same manner.
d. All children can learn, although not all children learn in the same manner.
e. All children can learn . . . although . . . not all children learn in the same manner.

31. If teachers set high expectations for there students, the students will rise to that high level.

a. there students
b. they're students
c. their students
d. his students
e. her students

32. In the modern age of technology, a teacher's focus is no longer the "what" of the content, but more importantly, the 'why.'

a. but more importantly, the 'why.'
b. but more importantly, the "why."
c. but more importantly, the 'why'.
d. but more importantly, the "why".
e. but more importantly the why.

33. Students have to read between the lines, identify bias, and determine who they can trust in the milieu of ads, data, and texts presented to them.

a. read between the lines, identify bias, and determine
b. read between the lines, identify bias, and determining
c. read between the lines, identifying bias, and determining
d. reads between the lines, identifies bias, and determines
e. reading between the lines, identify bias, and determine

34. During their time in present-day Newfoundland, Leif's expedition made contact with the natives whom they referred to as Skraelings (which translates to 'wretched ones' in Norse).

a. (which translates to 'wretched ones' in Norse).
b. (which translates to "wretched ones" in Norse.)
c. (which translates to 'wretched ones' in Norse.)
d. (which translates to "wretched ones" in Norse).
e. which translates to (wretched ones) in Norse.

35. Above all, it allowed us to share adventures. While traveling across America, which we could not have experienced in cars and hotels.

a. Above all, it allowed us to share adventures. While traveling across America
b. Above all, it allowed us to share adventures while traveling across America
c. Above all, it allowed us to share adventures; while traveling across America
d. Above all, it allowed us to share adventures—while traveling across America
e. Above all it allowed us to share adventures while traveling across America

36. Which is the best version of the underlined portion of this sentence (reproduced below)?

<u>Those are also memories that my siblings and me</u> have now shared with our own children.

a. Those are also memories that my siblings and me
b. Those are also memories that me and my siblings
c. Those are also memories that my siblings and I
d. Those are also memories that I and my siblings
e. Those are also memories that us siblings and I

Answer Explanations #2

Integrated Reasoning

Multi-Source Reasoning

1. NO: The employees in the previous year pay $500 monthly, which totals $6,000 per year. Next year, the employees will pay $250 biweekly, which totals $6,500.

NO: The other plans might make more sense for some employees who anticipate having high medical bills.

YES: The low deductible plan costs $200 biweekly, which totals $5,200 per year.

2. NO: The high deductible plan will cost the employee $6,800 ($1300 in premiums plus $5,000 toward the deductible and 10% of costs after deductible, which is $500). The middle deductible plan will cost the employee $6,650 ($2600 in premiums plus $3,000 toward the deductible and 15% of costs after deductible, which is $1,050).

YES: An employee who projects no medical costs should choose the plan with the lowest premium, and that is the high deductible plan, which costs the employee $1,300 a year.

YES: The high deductible plan will cost the employee $16,300, which is the lowest cost. This includes the $1,300 paid toward the premium, $5,000 toward the deductible, and 10% of $100,000, which equals $10,000 paid toward costs after the deductible.

3. A: The cost of tuition is increasing by 1% next year, which means that for an out-of-state resident, the cost per credit hour is $1,060.50. If a student takes 12 credit hours but then receives a 15% discount, the total tuition is 85% of $12,726, which is $10,817.10. This amount rounded to the nearest dollar is $10,817.

4. NO: Both students are taking fewer than 11 credit hours, so they both enroll on the third Monday in April.

NO: Students who take more than 7 credit hours next year will actually see their tuition decrease due to the new discounts.

YES: An in-state student taking 19 credit hours will pay $9,453.60 for the first 18 credit hours and $656.50 for the 19th credit hour, which totals $10,110.10

Graphic Interpretation

5. B: There are 15 employees out of 45 total employees that have worked at the company between 5 and 10 years. Therefore, the percentage is $\frac{15}{45} = 33.3\%$, rounded to the nearest tenth.

6. B: Because there are 267 seniors that are planning on going to school full time next year and that equals 53.9% of the total senior class, there are $267 \div 0.539 = 495$ total seniors at Heights High, rounded to the nearest integer. That means that 6.1% of 495 seniors are headed into the military, which equals $495 \times 0.061 = 30$, rounded.

7. B: The regression line predicts the output, given a specific input. Therefore, to approximate the miles driven in 7 hours, locate 7 on the x-axis, which corresponds to time. Then, locate the regression line at that x-value and find its corresponding y-value. Its corresponding y-value is approximately 57, which refers to 57 miles.

Two-Part Analysis

8. The correct value for x is 1296 and the correct value for y is 72 because $\frac{1296}{72} = 18$, which is a multiple of 9 because:

$$2 \times 9 = 18$$

9. There are 1,350 total business majors. Let *a* represent the number of accounting majors, and therefore the algebraic expression 3*a* can represent the number of marketing majors because there are 3 times as many accounting majors as marketing majors. There are 50 double majors. Therefore, we have the equation $1350 = a + 3a - 50$, which represents the total number of business majors. Subtracting the 50 represents any double-counting of double majors. Solving for *a* results in $a = 350$ the number of accounting majors. Therefore $3a = 1050$, the number of marketing majors. Therefore, the correct value for accounting is 350 and the correct value for marketing is 1,050.

Table Analysis

10. FALSE: The total number of basic packages sold was 135,800. However, dividing this amount by 4 results in an average of 33,950.

FALSE: The total number of non-basic packages sold was 251,100, and the total number of packages sold was 386,900, which is a percentage of 65%.

TRUE: The region with the largest number of internet sales for the ABC Cable Company was the Midwest. Totaling up the number of packages for each region results in the largest total for the Midwest region, 121,300 packages.

11. FALSE: The average number of credit hours completed was 15.3, which rounds to 15.

FALSE: Given the sample, the probability that a student is an engineering major is equal to the probability that a student is a physical therapy major because there were the same number of both majors in the sample.

FALSE: The data set corresponding to ACT scores has one mode, which is 31. It appears 3 times in the data set. No other score appears 3 times, so there is only one mode. A data set that is bimodal has 2 modes.

12. TRUE: Percent change is calculated by taking the ratio in change in population from year to year over the previous amount. For instance, from 2001 to 2002, the percent change is $\frac{301.2-300.1}{300.1} \times 100 = 0.366\%$, which rounds up to .37%. The population increased .37%. The year with the largest change occurred from 2002 to 2003, with a .43% decrease in population.

FALSE: The largest percent increase occurred from 2001 to 2002. The largest percent decrease occurred from 2002 to 2003.

TRUE: The population increased 0.38% from 2007 to 2008 and decreased 0.38% from 2004 to 2005.

Quantitative Reasoning

Problem Solving

1. B: To simplify the given equation, the first step is to make all exponents positive by moving them to the opposite place in the fraction. This expression becomes $\frac{4b^3b^2}{a^1a^4} \times \frac{3a}{b}$. Then the rules for exponents can be used to simplify. Multiplying the same bases means the exponents can be added. Dividing the same bases means the exponents are subtracted. Thus, after multiplying the exponents in the first fraction the equation becomes $\frac{4b^5}{a^5} \times \frac{3a}{b}$. Therefore, we can first multiply to get $\frac{12ab^5}{a^5b}$. Then, dividing yields $12\frac{b^4}{a^4}$.

2. E: The product of two irrational numbers can be rational or irrational. Sometimes, the irrational parts of the two numbers cancel each other out, leaving a rational number. For example, $\sqrt{2} \times \sqrt{2} = 2$ because the roots cancel each other out. Technically, the product of two irrational numbers can be complex because complex numbers can have either the real or imaginary part (in this case, the imaginary part) equal zero and still be considered a complex number. However, Choice *D* is incorrect because the product of two irrational numbers is not an imaginary number so saying the product is complex *and* imaginary is incorrect.

3. B: The car is traveling at a speed of five meters per second. On the interval from one to three seconds, the position changes by ten meters. By making this change in position over time into a rate, the speed becomes ten meters in two seconds or five meters in one second.

4. C: The number negative four is classified as a real number because it exists and is not imaginary. It is rational because it does not have a decimal that never ends. It is an integer because it does not have a fractional component. The next classification would be whole numbers, for which negative four does not qualify because it is negative. Choices *D* and *E* are wrong because -4 is not considered an irrational number because it does not have a never-ending decimal component.

5. E: The degree of freedom for two samples is calculated as $df = \frac{(n_1-1)+(n_2-1)}{2}$ rounded to the lowest whole number. For this example:

$$df = \frac{(29-1)+(30-1)}{2} = \frac{28+29}{2} = 28.5$$

which, rounded to the lowest whole number, is 28.

6. B: There are two zeros for the function $x = 0, -2$. The zeros can be found several ways, but this particular equation can be factored into:

$$f(x) = x(x^2 + 4x + 4) = x(x+2)(x+2)$$

By setting each factor equal to zero and solving for x, there are two solutions. On a graph, these zeros can be seen where the line crosses the x-axis.

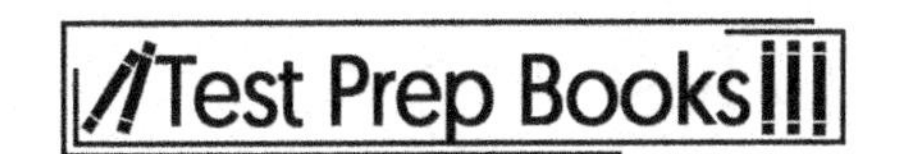

7. D: This problem involves a composition function, where one function is plugged into the other function. In this case, the $f(x)$ function is plugged into the $g(x)$ function for each x-value. The composition equation becomes:

$$g(f(x)) = 2^3 - 3(2^2) - 2(2) + 6$$

Simplifying the equation gives the answer:

$$g(f(x)) = 8 - 3(4) - 2(2) + 6$$

$$8 - 12 - 4 + 6 = -2$$

8. D: This system of equations involves one quadratic function and one linear function, as seen from the degree of each equation. One way to solve this is through substitution. Solving for y in the second equation yields:

$$y = x + 2$$

Plugging this equation in for the y of the quadratic equation yields:

$$x^2 - 2x + x + 2 = 8$$

Simplifying the equation, it becomes:

$$x^2 - x + 2 = 8$$

Setting this equal to zero and factoring, it becomes:

$$x^2 - x - 6 = 0 = (x - 3)(x + 2)$$

Solving these two factors for x gives the zeros $x = 3, -2$. To find the y-value for the point, each number can be plugged in to either original equation. Solving each one for y yields the points $(3, 5)$ and $(-2, 0)$.

9. D: The expression is simplified by collecting like terms. Terms with the same variable and exponent are like terms, and their coefficients can be added.

10. A: Finding the product means distributing one polynomial onto the other. Each term in the first must be multiplied by each term in the second. Then, like terms can be collected. Multiplying the factors yields the expression:

$$20x^3 + 4x^2 + 24x - 40x^2 - 8x - 48$$

Collecting like terms means adding the x^2 terms and adding the x terms. Then, simplify:

$$20x^3 - 36x^2 + 16x - 48$$

11. E: Finding the zeros for a function by factoring is done by setting the equation equal to zero, then completely factoring. Since there is a common x for each term in the provided equation, that should be factored out first. Then the quadratic that is left can be factored into two binomials, which are $(x + 1)(x - 4)$. Setting each factor equal to zero and solving for x yields three zeros.

12. D: Dividing rational expressions follows the same rule as dividing fractions. The division is changed to multiplication by the reciprocal of the second fraction. This turns the expression into:

$$\frac{5x^3}{3x^2y} \times \frac{3y^9}{25}$$

This can be simplified by finding common factors in the numerators and denominators of the two fractions.

$$\frac{x^3}{x^2y} \times \frac{y^9}{5}$$

Multiplying across creates:

$$\frac{x^3y^9}{5x^2y}$$

Simplifying leads to the final expression is:

$$\frac{xy^8}{5}$$

13. B: The equation can be solved by factoring the numerator into $(x+6)(x-5)$. Since that same factor exists on top and bottom, that factor $(x-5)$ cancels. This leaves the equation:

$$x + 6 = 11$$

Solving the equation gives the answer $x = 5$. When this value is plugged into the equation, it yields a zero in the denominator of the fraction. Since this is undefined, there is no solution.

14. D: This problem can be solved by using unit conversion. The initial units are miles per minute. The final units need to be feet per second. Converting miles to feet uses the equivalence statement 1 mile equals 5,280 feet. Converting minutes to seconds uses the equivalence statement 1 minute equals 60 seconds. Setting up the ratios to convert the units is shown in the following equation:

$$\frac{72 \text{ mi}}{90 \text{ min}} \times \frac{1 \text{ min}}{60 \text{ s}} \times \frac{5280 \text{ ft}}{1 \text{ mi}} = 70.4\frac{\text{ft}}{\text{s}}$$

The initial units cancel out, and the new units are left.

15. B: The formula can be manipulated by dividing both the length, *l*, and the width, *w*, on both sides. The length and width will cancel on the right, leaving height by itself.

16. B: The domain is all possible input values, or *x*-values. For this equation, the domain is every number greater than or equal to zero. There are no negative numbers in the domain because taking the square root of a negative number results in an imaginary number.

17. E: This problem can be solved by setting up a proportion involving the given information and the unknown value. The proportion is:

$$\frac{21 \text{ pages}}{4 \text{ nights}} = \frac{140 \text{ pages}}{x \text{ nights}}$$

Solving the proportion by cross-multiplying, the equation becomes $21x = 4 \times 140$, where $x = 26.67$. Since it is not an exact number of nights, the answer is rounded up to 27 nights. Twenty-six nights would not give Sarah enough time.

18. D: The slope from this equation is 50, and it is interpreted as the cost per gigabyte used. Since the g-value represents number of gigabytes and the equation is set equal to the cost in dollars, the slope relates these two values. For every gigabyte used on the phone, the bill goes up 50 dollars.

19. C: Graphing the function $y = \cos(x)$ shows that the curve starts at $(0, 1)$, has an amplitude of 2, and a period of 2π. This same curve can be constructed using the sine graph, by shifting the graph to the left $\frac{\pi}{2}$ units. This equation is in the form:

$$y = \sin\left(x + \frac{\pi}{2}\right)$$

20. D: Using SOHCAHTOA, tangent is $\frac{y}{x}$ for the special triangles. Since the value of $\tan(x)$ needs to be negative one, the angle for the tangent must be some form of 45 degrees or $\frac{\pi}{4}$. The value is negative in the second and fourth quadrant, so the answer is $\frac{3\pi}{4}$ and $\frac{7\pi}{4}$.

Data Sufficiency

21. B: Statement (1) is insufficient. We need to know how many packages there are total to find the number of potential selections. (A) and (D) are not possibilities. Statement (2) is sufficient. If there are 10 packages available and 4 possible selections, there are:

$$\binom{10}{4} = \frac{10 \times 9 \times 8 \times 7}{4 \times 3 \times 2} = 210 \text{ combinations}$$

22. A: Statement (1) states that Katie can finish the story in 4 hours, which means that Serena can finish the story in 8 hours because Katie is twice as fast as Serena. Because both of these quantities create rates, together they can finish one story in:

$$\frac{1}{4} + \frac{1}{8} = \frac{2}{8} + \frac{1}{8} = \frac{3}{8}$$

Dividing 8 by 3 shows that together they can finish the story in $2\frac{2}{3}$ hours. Therefore (1) is sufficient. Statement (2) is insufficient because we do not know how long the story is that they need to complete.

23. B: Statement (1) is not sufficient because it gives the following solution for y: $(-\infty, -5) \cup (5, \infty)$. That means that y could be negative if it exists in the first interval. Solving statement (2) results in $y > 4$, which is the interval $(4, \infty)$. Therefore, this shows that y must be positive. Statement (2) is sufficient.

24. D: If $y = z$, then the ratio $\frac{y}{z} = 1$, so statement (1) shows that $y \neq z$. Solving statement (2) results in the inequality $z < y - 2$, so this also shows that $y \neq z$. Therefore, each statement alone is sufficient to show that the answer is no.

25. B: Statement (1) is insufficient because we still need to know what c is in order to solve for a. Statement (2) gives the result that the sum of b and c is equal to 120. Plugging this into the equation results in $a + 120 = 160$, which means that $a = 40$. Therefore, statement (2) is sufficient by itself.

26. B: Statement (1) is insufficient. If x is a multiple of 6, it could be equal to 24. $9 - \frac{24}{9}$ does not result in an integer. Not all multiples of 6 are divisible by 9. Statement (2) is sufficient. Because $\frac{x}{9}$ has a remainder of 0, that means x is a multiple of 9 and the division results in an integer. An integer subtracted from 9, which is an integer, results in an integer.

27. D: Statement (1) is sufficient. Because there is a 25% profit on each box, oatmeal is sold for:

$$\$2 + 0.25 \times \$2 = \$2 + \$0.50 = \$2.50$$

Statement (2) is also sufficient. Because your selling price is 80% of your purchase price, $\$2 = 0.8 \times p$, where p is the purchase price. Therefore, $p = \frac{\$2}{.8} = \$2.50$. Both statements (1) and (2) are sufficient individually.

28. E: Neither statement (1) nor statement (2) tells us the shape of the pantry. It could be rectangular or irregularly shaped. Therefore, neither one gives us enough information to answer the question. We need to know the exact dimensions of the walls of the pantry to calculate its perimeter.

29. B: Statement (1) is not sufficient. If $x < 1$, it could be a fraction between 0 and 1 and in this case, $x^4 < x$. However, if $x < -1$, which is part of this interval, $x^4 > x$. Statement (2) is sufficient by itself. If $x < -2$ then $x^4 > x$. The answer to this question is no.

30. E: Statement (1) is not sufficient. If the mean is 3, then the sum of values is 21. If the median is 3, the middle of the values is 3. Neither one answers the question. Statement (2) is not sufficient. Knowing how many times the mode appears does not help us determine whether the median is less than the mode.

31. C: Statement (1) is insufficient by itself because it does not give us the height or radius of the cylinder. Statement (2) tells us that the base is 77 square centimeters, so its base is 7 by 11 centimeters. Therefore, the height is the third dimension: 12 centimeters. This is enough information to obtain the radius from the surface area given in (1). The height and radius are necessary to determine the volume of the cylinder. Therefore, both statements are sufficient together.

32. C: Statement (1) tells us one point on the line, but we need 2 points to find the slope of a line in order to determine the equation of the line. The second statement gives us that information. Together, they tell us that the slope is -2 and the equation of the line is $y = 2x + 6$, which (-2, 10) is on. The answer is yes.

33. C: In order for 2 lines to be perpendicular, the product of their slopes must equal -1. Statement (1) is insufficient by itself because it tells us one slope, 5. Statement (2) is insufficient by itself because it tells us the other slope. Putting the equation in slope-intercept form gives us $y = -\frac{1}{5}x + 2$, so its slope is $-\frac{1}{5}$. Both statements individually tell us a slope of one of the lines, so we can see that the lines are perpendicular because the product of the slopes is equal to -1.

34. C: Statement (1) is insufficient because we need to know how many red marbles there are and it tells us that there are 14 blue marbles. Statement (2) tells us that the probability of selecting a green marble is equal to the probability of selecting a blue marble. Therefore, the number of blue marbles equals the number of green marbles, which is given to us as 14 in statement 1. Because there are 39

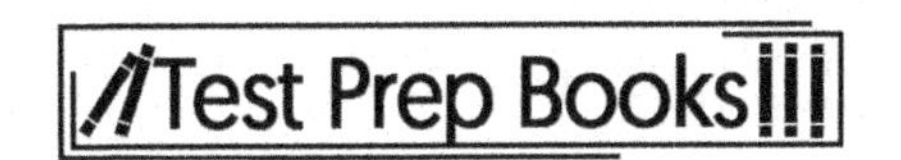

total marbles, that means 11 are red. Together, both statements enable us to find the probability of selecting a red marble as $11/39$.

35. D: Statement (1) gives us enough information to calculate the standard deviation as 3.7 because it is equal to the difference between the mean and the first data point below the mean, which is one standard deviation away. Statement (2) also tells us that the standard deviation is 3.7 because the difference between the data point 2 standard deviations from the mean and the data point one standard deviation from the mean is equal to the standard deviation.

36. A: From simplification, this rational expression reduces to:

$$\frac{-a^6(a-b)^2}{(d-b)^3}$$

The numerator is negative because $-a^6 > 0$ and $(a-b)^2 > 0$. Anything with an even exponent is nonnegative, and we know also that $a \neq b$. Therefore, we need to know the sign of the denominator. Statement (1) is tells us that $b > d$. Therefore, $b - d > 0$ and $0 > d - b$. Therefore, the denominator is negative because a negative value cubed is negative. The entire expression is positive. Statement (2) is insufficient.

37. E: Both positive and negative values must be considered. Statement (1) tells us that $|x| > |y|$. Two scenarios could be $x = 4, y = 2$ or $x = -4, y = -2$. In this case, they both give different answers to the question. Therefore, statement (1) is insufficient by itself. Statement (2) gives a similar argument. Because $a = 4b$, 2 scenarios could be $a = 4, b = 1$ and $a = -4,\ \ b = -1$. Both scenarios give different answers to the question. Therefore statement (2) is insufficient by itself. The statements together are still insufficient. The scenarios $a\ = 4, b = 1$ and $a = -4,\ \ b = -1$ satisfy both statements but still give an inconclusive answer.

Verbal Reasoning

Reading Comprehension

1. D: The passage references statements made by the World Bank in 1995 at the end of the first paragraph and in the middle of the second. The first reference says that state-owned enterprises "remain an important obstacle to better economic performance," and the second reference says the "World Bank recommended that SOEs be corporatized under commercial law and issued guidance." Thus, Choice *A* is incorrect and Choice *D* is correct. Although the World Bank is probably discussing developing countries, they are not mentioned in the statements quoted in the passage, so Choice *B* is incorrect. Choice *C* is contradicted by the rest of the passage and never mentioned in connection with statements issued by the World Bank. While the passage discusses studies similar to what's described in Choice *E*, they are not included in the World Bank's statements.

2. B: The passage repeatedly mentions disputes over privatization, including inconclusive studies, scholars refuting the World Bank's statements about state-owned enterprises, and differences between free market advocates who want to "unleash" the private sector and the World Bank's more gradual approach. Thus, Choice *B* is the correct answer. Choice *A* is incorrect because "always" is too strong. The studies are inconclusive and have yielded conflicting results. Choice *C* is incorrect for similar reasons. Although the studies in this passage are criticized, it's too much to say economic studies in general are always subject to such criticism. The World Bank recommends that state-owned enterprises be privatized in accordance with commercial law, but that doesn't necessarily mean those enterprises violate commercial law, so Choice *D* is incorrect. Nowhere in the passage does it say the World Bank holds the power to intervene in economies; its statements are referred to as recommendations. Thus, Choice *E* is incorrect.

3. E: World Bank includes competition and corporate governance free of political interference in its five preconditions for successful privatization, making Choice *E* the correct answer. The passage states that state-owned enterprises have a "pervasive presence" in the global economy, making Choice *D* incorrect. The empirical research is inconclusive but, if anything, it leans toward the opposite of what's described in Choice *A*. Influences on the World Bank's policies and popular support are never mentioned in the passage; therefore, Choices *B* and *C* are incorrect.

4. C: The author's conclusion is that "SOEs are an entity to be considered and managed in the pursuit of stability." As such, it can be inferred that the author supports the World Bank's measured approach of implementing privatization gradually to avoid the type of hasty action advocated by free market enthusiasts. Thus, Choice *C* is the correct answer. The author believes the market is the best allocator of resources, but it's unclear whether the author thinks private enterprises will always outperform state-owned enterprises, so this can't be the conclusion. Thus, Choice *A* is incorrect. The author would agree with Choices *B*, *D*, and *E*; however, all three are incorrect because they don't reflect the author's emphasis on stability.

5. C: The author mentions in the third paragraph how apologia focuses on the short term and reconciliation has long-term goals. Thus, Choice *C* is correct. Choice *D* is incorrect because reconciliation seeks to open channels of communication, not close them. The first clause in Choice *A* describes reconciliation rather than apologia, so it's incorrect. Choice *B* is incorrect because it's never stated or implied that apologia is always delivered in writing. The author doesn't claim that apologia requires actually receiving forgiveness to be effective, and reconciliation involves more than just delivering visual responses, so Choice *E* is incorrect.

6. E: The author states that national leaders use apologia and reconciliation to restore their image, so it can be inferred that a positive national image is an important part of governance. Thus, Choice *E* is the correct answer. The author clearly favors reconciliation, but it's unlikely they would agree that national leaders should never apologize. Thus, Choice *A* is incorrect. Similarly, the author implies that apologia is self-serving but, as described by the author, reconciliation also seems to be self-serving. In any event, the author thinks reconciliation is generally more effective because of its long-term effect, not because it's less self-serving. Thus, Choice *B* is incorrect. Choices *C* and *D* are never mentioned or alluded to in the passage, so they cannot be properly inferred.

7. B: Reconciliation's use of symbolic visual images and impact on the long-term relationship between the aggrieved and the transgressor is described in the third paragraph. Thus, Choice *B* is the correct answer. Choice A is incorrect because reconciliation involves long term, not short term, repair. The other answer choices describe features of apologia, not reconciliation, so Choices *C*, *D*, and *E* are all incorrect.

8. E: The author states in the fourth paragraph that offending images don't always require a visual response, but when they do, reconciliation is the better choice due to its long-term goals. Thus, Choice *E* is the correct answer. Censorship and anti-propaganda campaigns are never mentioned in the passage, so Choices *A* and *B* are incorrect. Choices *C* and *D* are incorrect because they are directly contradicted in the fourth paragraph.

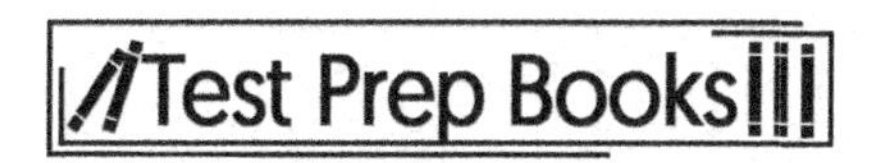

9. C: The first sentence of the second paragraph states, "While spending more than other countries per capita on health care services, the United States spends less on average than do other nations on social services impacting social and behavioral determinants of health." Thus, Choice *A* is incorrect and *C* is correct. Choice *B* is incorrect because it reverses the United States' ratio. The author argues that America's lower life expectancy is due to lower spending on social services, not insufficient funding for health care, so Choice *D* is incorrect. Choice *E* is incorrect because the United States does fund interdisciplinary programs that include behavioral and social science, like the Clinical and Translational Science Awards.

10. B: The Clinical and Translational Science Awards are mentioned in the third paragraph, and the author states that its interdisciplinary programs "aim to advance the translation of research findings from 'bench' to 'bedside' to 'community.'" Thus, Choice *B* is the correct answer. The Clinical and Translational Science Awards would likely support expanding social services, but advocating for that position isn't their purpose, so Choice *A* is incorrect. Similarly, the Clinical and Translational Science Awards likely conduct research on behavioral risks, but that's only part of the larger goal. So, Choice *C* is incorrect. Choice *D* is incorrect because employing social and behavioral scientists is not the Clinical and Translational Science Awards' primary purpose. It's unclear whether the Clinical and Translational Science Awards would even be involved in policy questions like calculating the optimal ratio for government spending, so Choice *E* isn't the primary purpose.

11. C: The author repeatedly mentions how the United States neglects social services. According to the author, this is why the United States performs poorly on key health indicators despite spending more per capita on health care than any other country. In addition, the author argues that the Clinical and Translational Science Awards should hire more social and behavioral scientists. Thus, Choice *C* is the correct answer. Choices *A* and *D* are premises that support the conclusion, not the main thesis, so they are both incorrect. Choice *B* is incorrect because it's only providing background information about the Clinical and Translational Science Awards. The author never asserts that life expectancy is the most important health care indicator, and, even if that were true, it wouldn't be the main thesis. Thus, Choice *E* is incorrect.

12. A: The purpose is to inform the reader about what assault is and how it is committed. Choice *B* is incorrect because the passage does not state that assault is a lesser form of lethal force, only that an assault can use lethal force, or alternatively, lethal force can be utilized to counter a dangerous assault. Choices *C* and *D* are incorrect because the passage is informative and does not have a set agenda. Finally, Choice *E* is incorrect because although the author uses an example in order to explain assault, it is not indicated that this is the author's personal account.

13. B: As discussed in the second passage, there are several forms of assault, like assault with a deadly weapon, verbal assault, or threatening posture or language. Choice *A* is incorrect because lethal force and assault are separate as indicated by the passages. Choice *C* is incorrect because anyone is capable of assault; the author does not state that one group of people cannot commit assault. Choice *D* is incorrect because assault is never justified. Self-defense resulting in lethal force can be justified. Choice *E* is incorrect because the author does mention what the charges are on assaults; therefore, we cannot assume that they are more or less than unnecessary use of force charges.

14. A: Both passages open by defining a legal concept and then describing situations in order to further explain the concept. Choice *D* is incorrect because while the passages utilize examples to help explain the concepts discussed, the author doesn't indicate that they are specific court cases. It's also clear that the passages don't open with examples, but instead, begin by defining the terms addressed in each passage. This eliminates Choice *B* and ultimately reveals Choice *A* to be the correct answer. Choice *A* accurately outlines the way both passages are structured. Because the passages follow a near identical structure, the rest of the choices can easily be ruled out.

Critical Reasoning

15. C: Choice *A* is not present in the teacher's argument. There is no discussion of a common rule, so this cannot be the answer. Eliminate this choice.

Choice *B* is inaccurate. The teacher justifies his argument by pointing to the two highest achieving students in his class. It's not a perfect argument, but it's untrue to say that the argument is totally unjustified.

Choice *C* looks much more promising than the other options. The teacher's conclusion is that it's *always* the case that students can overcome parental indifference. He supports this notion by pointing to the two best students in a class of twenty people. The conclusion is too broad when considering the evidence. This is probably the answer, but look at the other choices to make sure.

Choice *D* does not correspond with anything in the teacher's argument. Although the teacher doesn't present any counter arguments, the existence of a competing theory is unclear. Eliminate this choice.

Choice *E* is clearly incorrect. The argument does not show any bias. Eliminate this choice.

Therefore, Choice *C* is the correct answer.

16. C: Choice *A* restates a premise and does not resolve the paradox. Whether Trent's SWAT team is the best police unit does not answer why unsolved crimes are increasing every year despite historic rates of crime solving. Eliminate this choice.

Choice *B* also does not resolve the paradox. It attempts to dismiss the increase in unsolved crimes by characterizing those crimes as petty drug offenses. Even if true, this does not resolve the paradox.

Choice *C* is a strong answer choice. If the raw number of crimes increases every year, then it makes sense that crimes are increasing despite the historic rates of crime solving. This explains the paradox. In questions involving percentages, always pay special attention to answer choices that involve raw numbers.

Choice *D* is irrelevant. The competence of the police department does not explain the paradox. Additionally, the police department is apparently not incompetent since it's solving a higher percentage of crimes than ever before in its history. Eliminate this choice.

Choice *E* is similar to Choice *B*. It attempts to dismiss the increasing number of unsolved crimes by claiming that the police are solving the most important crimes. This does not explain the paradox.

Therefore, Choice *C* is the correct answer.

17. E: Choice *A* is unrelated to the argument's main point. This choice is misleading with extraneous information because the FDA is often criticized for this very reason. However, the argument does not address this point. There's no way it's the main point. Eliminate this choice.

Choice *B* is a very strong answer. The argument is definitely trending in this direction, especially since the argument points out the lack of tar in electronic cigarettes. Choice *B*'s use of *probably* fits with the argument's tone. Leave this option for now.

Choice *C* goes too far. The scientist's argument is more informational than directional. Choice *C* fails to match this tone. Eliminate this choice.

Choice *D* accurately restates one of the argument's premises, but it is not the main point. Eliminate this choice.

Choice *E* is an excellent balance of information and speculation, like the argument. The answer choice's first phrase identifies the concerns highlighted by the scientist, and the second phrase expresses why the scientist believes that electronic cigarettes are a promising alternative. Choice *B* is extremely similar, but Choice *E* better expresses the argument's main point.

Therefore, Choice *E* is the correct answer.

18. B: Choice *A* does not necessarily follow from the argument. Although Brittany's dog loves playing fetch, there's nothing in the argument that makes this definitely true. Eliminate this choice.

Choice *B* follows logically from the argument: *Only German shepherds love protecting their homes.* In other words, no other dogs love protecting their homes. Therefore, if Brittany's dog loves protecting her home, then it must be a German shepherd.

Choice *C* is clearly incorrect. Just because some dogs are easy to train does not mean that Brittany's dog is easy to train. Eliminate this choice.

Choice *D* is tricky but incorrect. Although the last sentence references both qualities attributed to Labrador retrievers and German shepherds, there is no information concerning a mix of the two. Eliminate this choice.

Choice *E* references all of the information included in the argument, but it doesn't follow logically. There is nothing in the argument that suggests that Choice *E* must be true. Eliminate this choice.

Therefore, Choice *B* is the correct answer.

19. E: Choice *A* supports the argument. The argument concludes that poverty is the number one cause of crime. It would make sense that criminals are less wealthy than the average person. Keep this choice for now.

Choice *B* is irrelevant for the purposes of this argument. Even if redistributing wealth is indeed the best way to lessen poverty, it does not support the connection between poverty and crime. Eliminate this choice.

Choice *C* is tangentially related to the argument. If substance abuse is the second largest cause of crime and those abusers are poor, then it makes sense that criminals are poor. However, this answer choice is worse than Choice *A*, which explicitly states the same thing. Eliminate this choice.

Choice *D* is irrelevant. The argument makes no mention of morality or how moral societies would treat the poor. Eliminate this choice.

Choice *E* is a very strong answer. If the majority of crimes involve food theft and trespassing, then it supports the notion that people commit crimes to meet their basic needs. Choice *E* strengthens the conclusion that if these people's basic needs were met then the majority of crime would not be committed. This offers more support than Choice *A*.

Therefore, Choice *E* is the correct answer.

20. A: Choice *A* is a strong answer choice. Negate the choice to see if it's a necessary assumption: *Strength and size gains are NOT indicators of good health.* This destroys the argument. If strength and size are not indicators of good health, then regular weightlifting is not necessary for good health. This is probably the correct answer, but work through the remaining options.

Choice *B* is not a necessary assumption. The argument is no worse off if there are other ways to increase strength and size besides compound movements. Eliminate this choice.

Choice *C* is not a contention made by the argument. The argument merely states that compound movements are an especially effective type of heavy resistance weightlifting. This is definitely not a necessary assumption. Eliminate this choice.

Choice *D* is also clearly incorrect. Don't be fooled by the parallels with Choice *C*. This choice is wrong for the same reason as Choice *C*. Eliminate this choice.

Choice *E* restates the conclusion, so it is not a necessary assumption. It is an explicit conclusion. Eliminate this choice.

Therefore, Choice *A* is the correct answer.

21. C: Choice *A* definitely strengthens the argument by highlighting a benefit of an autocratic government. Eliminate this choice.

Choice *B* seems to strengthen the argument. This answer choice connects the start of the autocratic despot's reign with the start of economic growth. Eliminate this choice.

Choice *C* appears to weaken the argument. This answer choice provides an alternate explanation for the economic growth. According to Choice *C*, West Korea experienced economic growth as a result of the oil reserve. This hurts the argument's contention that West Korea's economy benefits from limiting civil liberties. This is a very strong answer choice.

Choice *D* clearly strengthens the argument. If political protest harms economic growth, then there's additional support for West Korea's curtailment of civil liberties. Eliminate this choice.

Choice *E* also strengthens the argument. The despot is able to devote all of his time to solving economic problems since there are no civil liberties. Eliminate this choice.

Therefore, Choice *C* is the correct answer.

22. E: Choice *A* is too extreme. The sociologist definitely believes that the abolition of marriage would harm society, but collapse goes too far. Eliminate this choice.

Choice *B* is tricky since the previous sentence references how children born out of wedlock are more likely to work at lower paying jobs. However, this answer choice also goes too far. According to the

argument, unmarried people are less likely to be homeowners or save for retirement. But if marriage rates decline, it is not necessarily true that everyone would have less money. Eliminate this choice.

Choice *C* is tricky but also incorrect. Unmarried people are less likely to own homes. If marriage rates decline, fewer people would be homeowners. This is not the same as arguing that nobody would own homes. Eliminate this choice.

Choice *D* is irrelevant to the argument. The argument makes no reference to happiness. There is no way that such new information would logically complete the passage. Eliminate this choice.

Choice *E* looks much more promising than the other choices. If unmarried people are less likely to attend college and marriage rates decline, then it is reasonable to say that college attendance would probably decrease. Choice *E* also matches the argument's tone through the use of *probably*, unlike many of the other answer choices.

Therefore, Choice *E* is the correct answer.

23. B: Choice *A* is not a necessary assumption. This answer choice provides additional support to the argument, but it is not dependent on this fact. The test developers hope the test takers will mistake this for a strengthening question. Don't be fooled. Eliminate this choice.

Choice *B* looks very promising. Negate this answer choice to see if the argument falls apart: *Consumers' disposable income is NOT directly related to their ability to purchase goods and services.* This hurts the argument. If disposable income is unrelated to purchasing goods and services, then tax rates don't matter. Definitely keep this answer.

Choice *C* is irrelevant to the argument. The argument does not depend on consumers' preferences. Eliminate this choice.

Choice *D* is a strong answer choice. Negate this answer choice to see if the argument falls apart: *Increasing disposable income is NOT the only way to ensure economic growth.* The argument is definitely worse off, but it is not destroyed. Therefore, this is not a necessary assumption. Eliminate this choice.

Choice *E* is irrelevant to the argument. The argument discusses how tax rates impact economic growth. It does not mention social welfare programs. Always be careful of new information, like the role social welfare programs play in this choice. Eliminate this choice.

Therefore, Choice *B* is the correct answer.

24. B: Choice *A* is similar to the argument in that it makes a prediction based on past events; however, Choice *A*'s argument is much more reasonable than the argument. If Ted has eaten an apple pie every day for the last decade, then it's reasonable to assume that he will do so again tomorrow. Eliminate this choice.

Choice *B* is a very strong answer choice. Like the argument, it takes past events and speculates that conditions will not change. Just as any number of factors could alter the United States' economic growth, it is similarly unreasonable to say that Alexandra will be the top salesperson based on one year's data. This answer choice also uses extremely strong language in its speculation (*guaranteed* and *undoubtedly*). Definitely keep this answer choice as an option.

Choice *C* is similar to the argument, but its conclusion is much more reasonable. If George has brushed his teeth right before bed for twenty years, then it is not unreasonable to speculate that he will do the same tonight. This is not the same as predicting that past economic conditions will continue into the future. George has much more control over brushing his teeth than the United States has over its economy. Eliminate this choice.

Choice *D* mirrors the language as the argument, but it draws a very different conclusion. In contrast to the argument, Choice *D*'s conclusion gives a reason why Germany's economy is on the rise. It does not make a guarantee of future growth. This is not the same as the argument. Eliminate this choice.

Choice *E* does not rely on flawed reasoning, so it must be incorrect. If Tito is the top ranked surfer in the world and listed as a clear favorite, then it's true that he's the most likely to win the tournament. Eliminate this choice.

Therefore, Choice *B* is the correct answer.

25. B: Choice *A* weakens one of the argument's premises. If big cats don't try to escape because they can't figure out their enclosures, then never attempting to escape is not a sign of intelligence. This definitely weakens the argument by negating one of its premises. Keep it for now.

Choice *B* looks extremely promising. This answer choice tells us that experts disagree that adjusting to captivity is a measure of intelligence. If big cats' adjustment to captivity does not correspond to intelligence, then the zookeeper's entire argument is flawed. This destroys the argument.

Choice *C* weakens the argument, but it's less powerful than Choice *B*. If bears share similarities with big cats, then there might be some doubt as to which animal is the smartest land mammal. This weakens the argument, but not as much as Choice *B*, which completely disrupts the argument's logic. Eliminate this choice.

Choice *D* actually strengthens the zookeeper's argument. The brain scans support the zookeeper's conclusion that big cats are the smartest land mammals. Eliminate this choice.

Choice *E* is a strong answer choice. If the zoo is devoting significantly more resources to caring for big cats, then the difference in resources could be the reason for their adaptability. However, Choice *B* spoils the argument's entire logical thrust. Eliminate Choice *E*.

Therefore, Choice *B* is the correct answer.

26. C: Choice *A* is incorrect. The argument states that nearly all lawyers dutifully represent their clients' best interest. *Nearly all* is not the same as *all*. It can't be definitively said that it must be true that Tanya represents her clients' best interests. Eliminate this choice.

Choice *B* is incorrect. The argument states that only some lawyers charge exorbitant and fraudulent fees. Thus, Tanya is not necessarily one of these bad apple attorneys. Eliminate it.

Choice *C* follows the argument's reasoning. The argument states that all lawyers are bound by extensive ethical codes. Therefore, if Tanya is a lawyer, then she must be bound by extensive ethical codes. This is the correct answer.

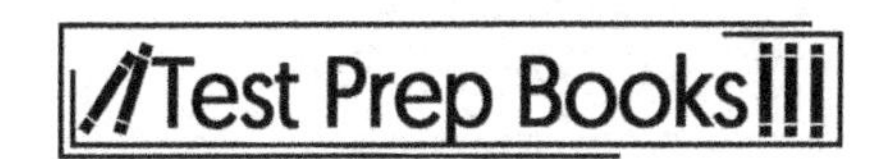

Choice *D* is incorrect. The argument states that only some lawyers become millionaires. Therefore, it's not necessarily true that Tanya is a millionaire. She could be, but it doesn't have to be true. Eliminate this choice.

Choice *E* is incorrect. Similar to Choice *D*, the argument states that only some lawyers work in the public sector. Therefore, it's not necessarily true. Eliminate this choice.

Therefore, Choice *C* is the correct answer.

Sentence Correction

27. B: Choice *B* is correct. Here, a colon is used to introduce an explanation. Colons either introduce explanations or lists. Additionally, the quote ends with the punctuation inside the quotes, unlike Choice *C*.

28. D: The word *civilised* should be spelled *civilized.* The words "makes," "human," "distinguishes," and "creatures," are all spelled correctly.

29. A: Choice *A* is correct because the phrase "others' ideas" is both plural and indicates possession. Choice *B* is incorrect because "other's" indicates only one "other" that's in possession of "ideas," which is incorrect. Choice *C* is incorrect because no possession is indicated. Choice *D* is incorrect because the word "other" does not end in *s*. Others's is not a correct form of the plural possessive word. Choice *E* is incorrect, as there is no reason the word "others" should be in quotations.

30. D: This sentence must have a comma before "although" because the word "although" is connecting two independent clauses. Thus, Choices *B* and *C* are incorrect. Choice *A* is incorrect because the second sentence in the underlined section is a fragment. Choice *E* is incorrect. Ellipses (. . .) are used to indicate an omitted phrase.

31. C: Choice *C* is the correct choice because the word "their" indicates possession, and the text is talking about "their students," or the students of someone. Choice *A*, "there," means at a certain place and is incorrect. Choice *B*, "they're," is a contraction and means "they are." Choices *D* and *E* are incorrect because these contain a singular pronoun, and our noun, "teachers," is plural.

32. B: Choice *B* uses all punctuation correctly in this sentence. In American English, single quotes should only be used if they are quotes within a quote, making Choices *A* and *C* incorrect. Additionally, punctuation should go inside quotation marks with a few exceptions, making Choice D incorrect. Choice *E* is incorrect; the "why" should be in some sort of quotes since it is a questioning term within the sentence.

33. A: Choice *A* has consistent parallel structure with the verbs "read," "identify," and "determine." Choices *B* and *C* have faulty parallel structure with the words "determining" and "identifying." Choice *D* has incorrect subject/verb agreement. The sentence should read, "Students have to read . . . identify . . . and determine." Choice *E* is incorrect; the word "reading" should be "read."

34. D: Choice *D* uses the correct punctuation. American English uses double quotes unless placing quotes within a quote (which would then require single quotes). Thus, Choices *A* and *C* are incorrect. Choice *B* is incorrect because the period should go outside of the parenthesis, not inside. Choice *E* is incorrect; parenthesis around "wretched ones" causes the sentence to have problems. A sentence should be able to stand on its own without the words in the parenthesis completing it.

35. B: Choice *B* is correct because there is no punctuation needed if a dependent clause ("while traveling across America") is located behind the independent clause ("it allowed us to share adventures"). Choice *A* is incorrect because there are two dependent clauses connected and no independent clause, and a complete sentence requires at least one independent clause. Choice *C* is incorrect because of the same reason as Choice *A*. Semicolons have the same function as periods: there must be an independent clause on either side of the semicolon. Choice *D* is incorrect because the dash simply interrupts the complete sentence. Choice *E* is incorrect because there is no comma after the introductory phrase "Above all."

36. C: The rules for "me" and "I" is that one should use "I" when it is the subject pronoun of a sentence, and "me" when it is the object pronoun of the sentence. Break the sentence up to see if "I" or "me" should be used. To say "Those are memories that I have now shared" is correct, rather than "Those are memories that me have now shared." Choice *D* is incorrect because "my siblings" should come before "I." Choice *E* is incorrect because the pronoun "us" should be "my."

Index

Dear GMAT Test Taker,

We would like to start by thanking you for purchasing this study guide for your ASWB Bachelors exam. We hope that we exceeded your expectations.

Our goal in creating this study guide was to cover all of the topics that you will see on the test. We also strove to make our practice questions as similar as possible to what you will encounter on test day. With that being said, if you found something that you feel was not up to your standards, please send us an email and let us know.

We would also like to let you know about other books in our catalog that may interest you.

GRE

This can be found on Amazon: amazon.com/dp/162845900X

MCAT

amazon.com/dp/1628456779

We have study guides in a wide variety of fields. If the one you are looking for isn't listed above, then try searching for it on Amazon or send us an email.

Thanks Again and Happy Testing!
Product Development Team
info@studyguideteam.com

FREE Test Taking Tips DVD Offer

To help us better serve you, we have developed a Test Taking Tips DVD that we would like to give you for FREE. **This DVD covers world-class test taking tips that you can use to be even more successful when you are taking your test.**

All that we ask is that you email us your feedback about your study guide. Please let us know what you thought about it – whether that is good, bad or indifferent.

To get your **FREE Test Taking Tips DVD**, email freedvd@studyguideteam.com with "FREE DVD" in the subject line and the following information in the body of the email:

a. The title of your study guide.

b. Your product rating on a scale of 1-5, with 5 being the highest rating.

c. Your feedback about the study guide. What did you think of it?

d. Your full name and shipping address to send your free DVD.

If you have any questions or concerns, please don't hesitate to contact us at freedvd@studyguideteam.com.

Thanks again!

Made in the USA
Las Vegas, NV
13 November 2021